Lecture Notes in Mathematics

Edited by A. Dold and B. Eckmann

596

Klaus Deimling

Ordinary Differential Equations in Banach Spaces

Springer-Verlag
Berlin · Heidelberg · New York 1977

Author
Klaus Deimling
Fachbereich 17
der Gesamthochschule
Warburger Straße 100
D-4790 Paderborn

Library of Congress Cataloging in Publication Data

Deimling, Klaus, 1943-
Ordinary differential equations in Banach spaces.

(Lecture notes in mathematics ; 596)
Bibliography: p.
Includes index.
1. Differential equations. 2. Banach spaces.
I. Title. II. Series: Lecture notes in mathematics (Berlin) ; 596.
QA3.L28 no. 596 [QA372] 510'.8s [515'352] 77-22408

AMS Subject Classifications (1970): 34G05, 34F05, 47H10, 47H15, 60J80, 65J05

ISBN 3-540-08260-3 Springer-Verlag Berlin · Heidelberg · New York
ISBN 0-387-08260-3 Springer-Verlag New York · Heidelberg · Berlin

Printed in Germany
Printing and binding: Beltz Offsetdruck, Hemsbach/Bergstr.
2141/3140-543210

Preface

These notes represent an expanded version of material prepared for a one-semester graduate level course on differerential equations held at the University of Kiel in 1975 . The aim has been to show more or less recent connections between differential equations and functional analysis without assuming too many prerequisites from both fields , and with main emphasis on countable systems of ordinary differential equations.

In the main text we present the basic ideas and results. Generalizations and references are given in the Remarks to each chapter. Results without reference to the bibliography are either trivial or new or at least not in the literature available to us. In order to prevent controversy on priority let us mention explicitely that with no reference do we claim to have found the original place of a certain result. A typical example of historical uncertainty is mentioned in Remark (i) to § 4 . The Remarks and § 8 should also be understood as a guide to further studies.

Having already finished § 1 - § 7 we have been informed by Prof. R. Redheffer (UCLA) that Martin's book [113] was in the being and Prof. R.H. Martin has been so kind to send me a copy. Hereafter, we were informed by Prof. M. Kwapisz (Gdansk) about the existence of the book [199] of Valeev/Zautykov on countable systems. Naturally, § 2 - § 5 and § 6 - § 7 overlap in several topics with [113] and [199] , respectively.

Finally, I want to thank Jan Prüß who read the entire manuscript and eliminated several mistakes, my wife Brigitte for typing the manuscript and Prof. A. Dold for the possibility to publish the manuscript in the Lecture Notes series.

Klaus Deimling
Paderborn, April 1977

Contents

Introduction

The first investigations of countable systems of ordinary differential equations date back to the origins of functional analysis around the last turn of the century. The linear systems had been studied in the new framework of Hilbert's bounded quadratic forms and the nonlinear systems by means of the already well known method of successive approximation. In course of time one was led to consider such systems in connection with concrete problems in natural sciences. Let us sketch three examples.

Example 1 (Fourier's Method). Let us consider the heat conduction problem

(1) $$u_t = u_{xx} + f(t,x,u,u_x) \quad \text{for} \quad t \geq 0 \, , \; x \in [0,\pi]$$

(2) $$u(0,x) = \alpha(x) \text{ in } [0,\pi] \, , \; u(t,0) = u(t,\pi) = 0 \text{ in } t \geq 0 \, ,$$

where $u(t,x)$ denotes the temperature at time t and place x in a rod of length π . Suppose that the initial temperature α has the Fourier expansion

$$\alpha(x) = \sum_{n \geq 1} c_n \sin nx \, , \text{ with } c_n = \frac{2}{\pi} \int_0^\pi \alpha(s)\sin(ns)ds \, .$$

Let us try to find a solution u of (1) , (2) in the form

$$u(t,x) = \sum_{k \geq 1} u_k(t)\sin kx \, .$$

If we insert these series into (1) , multiply the equation by $\sin(nx)$ and integrate over $[0,\pi]$ then we obtain formally the following countable system for the unknown coefficients $u_n(t)$

$$u_n' + n^2 u_n = f_n(t,u_1,u_2,\ldots) \quad \text{for} \quad n \geq 1 \, , \; t \geq 0 \, ,$$

where

$$f_n(t,u_1,u_2,\ldots) = \frac{2}{\pi} \int_0^\pi f(t,x,\sum_{k \geq 1} u_k(t)\sin kx, \sum_{k \geq 1} k u_k(t)\cos kx)\sin nx dx \, .$$

By (2) , we also have the initial conditions $u_n(0) = c_n$ for $n \geq 1$. Some references for such problems are given in Remark (iii) to § 7 .

Example 2 (Partial Discretization). Consider again equation (1) , but now for $t \geq 0$ and all $x \geq 0$, together with the side condition

(3) $$u(0,x) = \alpha(x) \quad \text{in} \quad x \geq 0 \; , \; u_x(t,0) = -\beta(t) \quad \text{in} \quad t \geq 0 \; .$$

For numerical purposes it is natural to consider the corresponding difference equations, obtained by discretization of one of the variables t , x at least. For example, let us take a discretization with respect to x only , i.e. a step size $h > 0$, the grid points $x_n = nh$ for $n = 0,1,\ldots$, $u_n(t) = u(t,x_n)$, $h^{-1}(u_{n+1}(t)-u_n(t))$ instead of $u_x(t,x_n)$ and $h^{-2}(u_{n+1}+u_{n-1}-2u_n)$ instead of $u_{xx}(t,x_n)$. Then, we obtain the countable system

$$u_n' = h^{-2}(u_{n+1}+u_{n-1}-2u_n) + f(t,nh,u_n,h^{-1}(u_{n+1}-u_n)) \quad \text{for} \quad n \geq 1 \; , \; t \geq 0$$

with the initial conditions $u_n(0) = \alpha(nh)$ and $u_0(t) = u_1(t) + h\beta(t)$. Once this problem is solved by $u^h(t) = (u_1(t),u_2(t),\ldots)$, one takes a suitable interpolation in the direction of x , to obtain "approximate" solutions $u^h(t,x)$ which hopefully do converge to a true solution of (1) , (3) as $h \to 0$. For this approach, usually called "longitudinal method of lines" , we give some references in Remark (v) to § 5 and Remark (viii) to § 6 .

Example 3 ("Branching processes"). In some mathematical models for the propagation of bacteria, the production of neutrons in chain reactions, the production of electrons and photons in cosmic showers, etc., one has a system S which at every time $t \geq 0$ is in one of the countable states $n = 1,2,\ldots$. Let $p_i(t) = \text{prob}\{S(t) = i\}$, the probability that S is in state i at time t . If $\text{prob}(\cdot|\cdot)$ denotes the conditional probability then we have

$$p_i(t+h) = \sum_{j\geq 1} \text{prob}(S(t+h) = i|S(t) = j)p_j(t) \quad .$$

Now, one assumes that $\text{prob}(S(t+h)=i|S(t)=j) = a_{ij}h + o(h)$ for $i \neq j$ and $= 1 - a_{jj}h + o(h)$ for $i = j$ as $h \to 0$, where $a_{ij} \geq 0$ and $\sum_{i\neq j} a_{ij} = a_{jj}$. With this assumption, we obtain formally the system

(4) $$p_i' = - a_{ii}p_i + \sum_{j\neq i} a_{ij}p_j \quad , \quad p_i(0) = c_i \quad \text{for} \quad i \geq 1$$

with nonnegative constants c_i such that $\sum_{i\geq 1} c_i = 1$. In the literature such systems may also be found under the headings "Denumerable Markov chains" , "Random Walk" , "Birth-Death Process" ,"Queuing Theory" . This example will be considered at various places in the following chapters and references to concrete applications are given in Remark (vii) to § 7 .

One way to attack existence, uniqueness, stability etc. for such denumerable systems is to pick a suitable Banach space X of sequences, e.g. l^1 or $\{x \in \mathbb{R}^{\mathbb{N}} : \sum_{i\geq 1} a_{ii}|x_i| < \infty\}$ in Example 3 , and to consider the system as one differential equation $x' = f(t,x)$ for X-valued functions on $t \geq 0$, with initial conditions $x(0) = c \in X$. Since one is led to study such problems also in Banach spaces other than sequence spaces, e.g. in connection with timedependent partial,stochastic ordinary or integro-differential equations, we have devoted § 1 - § 5 to the initial value problem

(5) $$x' = f(t,x) \quad , \quad x(0) = c$$

in an arbitrary Banach space. Concerning existence and uniqueness for (5) , we shall consider right hand sides f that belong to some of the main classes of nonlinear mappings between Banach spaces, intensively studied in "Nonlinear Functional Analysis" during recent years.
In § 6 and § 7 we investigate the denumerable system as one equation in $\mathbb{R}^{\mathbb{N}}$ with the product topology, or by means of truncation, i.e. we cut the infinite system at the N-th row and the N-th column, solve the $N \times N$ system by u^N for every $N \geq 1$ and look for conditions on the right hand sides f_i ensuring that u^N tends to a solution of the infinite system as $N \to \infty$.
In the main, we have restricted ourselves to existence, uniqueness and differential inequalities. Up to now, nearly nothing has been done concerning qualitative properties of solutions, neither within nor outside of these notes.

§ 1 Lipschitz type conditions

Let X be a Banach space over $\mathbb{R}$ or $\mathbb{C}$, $D \subset X$, $J = [0,a] \subset \mathbb{R}$, $f: J\times D \to X$ and $x_0 \in D$. We look for continuously differentiable functions

$$x: [0,\delta] \to D \quad \text{for some} \quad \delta \in (0,a]$$

such that

(1) $$x' = f(t,x) \quad \text{in } [0,\delta] \quad , x(0) = x_0 \quad ;$$

such a function x is called a (local) solution of (1) .

1. Existence and uniqueness.

The following facts may be proved as in the case $X = R^n$, for example by means of successive approximations.
If f is continuous and satisfies the Lipschitz condition

$$|f(t,x) - f(t,y)| \leq L|x-y|$$

then (1) has a unique solution on J , provided D = X . If D is the ball $\bar{K}_r(x_0) = \{x:|x-x_0| \leq r\}$ then (1) has a unique solution on $[0,\delta]$, where $\delta = \min\{a,r/M\}$ and $M = \sup\{|f(t,x)| : t \in J, x \in D\}$. If D is open, f is continuous and locally Lipschitz (i.e. to each $(t,x) \in J\times D$ there exist $\eta = \eta(t,x) > 0$ and $L = L(t,x) > 0$ and a neighborhood U_x of x such that $|f(s,u) - f(s,v)| \leq L|u-v|$ for $s \in J \cap [t,t+\eta]$ and $u,v \in U_x$) then (1) has a unique solution, defined either on the whole interval J or only on a subinterval $[0,\delta)$ which is maximal with respect to extension of solutions.

2. Approximate solutions.

By means of the simple results just mentioned it is easy to construct approximate solutions for (1) in case f is only continuous. At first, we show that such an f may be approximated by locally Lipschitz functions.

Lemma 1.1. Let X,Y be Banach spaces, $\Omega \subset X$ open and $f: \Omega \to Y$ continuous. Then, to each $\varepsilon > 0$ there exists $f_\varepsilon: \Omega \to Y$, locally Lipschitz and such that $\sup_\Omega |f(x)-f_\varepsilon(x)| \leq \varepsilon$.

Proof. Let $U_\varepsilon(x) = \{y \in \Omega : |f(y)-f(x)| < \varepsilon/2\}$. We have

$$\Omega = \bigcup_{x \in \Omega} U_\varepsilon(x)$$

and $U_\varepsilon(x)$ open. Let $\{V_\lambda : \lambda \in \Lambda\}$ be a locally finite refinement of $\{U_\varepsilon(x) : x \in \Omega\}$, i.e. an open cover of Ω such that each $x \in \Omega$ has a neighborhood $V(x)$ with $V(x) \cap V_\lambda \neq \emptyset$ only for finitely many $\lambda \in \Lambda$, and such that to each $\lambda \in \Lambda$ there exists $x \in \Omega$ with $V_\lambda \subset U_\varepsilon(x)$. Define $\alpha_\lambda: X \to \mathbb{R}$ by

$$\alpha_\lambda(x) = \begin{cases} 0 & \text{for } x \notin V_\lambda \\ \rho(x,\partial V_\lambda) & \text{for } x \in V_\lambda \end{cases} ,$$

where $\rho(x,A) = \inf\{|x-y| : y \in A\}$. Let

$$\phi_\lambda(x) = \Big[\sum_{\mu \in \Lambda} \alpha_\mu(x)\Big]^{-1} \alpha_\lambda(x) \quad \text{for } x \in \Omega .$$

Since α_λ is Lipschitz on X and $\{V_\lambda : \lambda \in \Lambda\}$ is locally finite, ϕ_λ is locally Lipschitz on Ω . For every $\lambda \in \Lambda$ we choose some $a_\lambda \in V_\lambda$ and we define

$$(*) \qquad f_\varepsilon(x) = \sum_{\lambda \in \Lambda} \phi_\lambda(x) f(a_\lambda) \quad \text{for } x \in \Omega .$$

We have f_ε locally Lipschitz in Ω and

$$|f_\varepsilon(x)-f(x)| = \Big|\sum_{\lambda \in \Lambda} \phi_\lambda(x)\{f(a_\lambda)-f(x)\}\Big|$$

$$\leq \sum_{\lambda \in \Lambda} \phi_\lambda(x) |f(a_\lambda)-f(x)| .$$

Now, suppose $\phi_\lambda(x) \neq 0$. Then, $x \in V_\lambda \subset U_\varepsilon(x_0)$ for some x_0 and $a_\lambda \in V_\lambda$, hence $|f(a_\lambda)-f(x)| \leq \varepsilon$ and therefore

$$|f_\varepsilon(x)-f(x)| \leq \varepsilon \sum_\lambda \phi_\lambda(x) = \varepsilon .$$

q.e.d.

The following result on continuous extensions of continuous mappings may be proved along similar lines; see Remark (i) .

Lemma 1.2. Let X,Y be Banach spaces, $\Omega \subset X$ closed and $f\colon \Omega \to Y$ continuous. Then there is a continuous extension $\tilde{f}\colon X \to Y$ of f such that $\tilde{f}(X) \subset \operatorname{conv} f(\Omega)$ (= convex hull of $f(\Omega)$) .

Now, it is easy to find approximate solutions for (1) . We have

Theorem 1.1. Let $J = [0,a]$, $D = \overline{K}_r(x_0) \subset X$, $f\colon J\times D \to X$ continuous and $|f(t,x)| \leq M$ on $J\times D$. Let $\varepsilon > 0$ and $a_\varepsilon = \min\{a, r/(M+\varepsilon)\}$. Then there exists a continuously differentiable function $x_\varepsilon\colon [0,a_\varepsilon] \to D$ such that

$$x_\varepsilon' = f(t,x_\varepsilon)+y_\varepsilon(t) \quad , \quad x_\varepsilon(0) = x_0 \quad \text{and} \quad |y_\varepsilon(t)| \leq \varepsilon \quad \text{on } [0,a_\varepsilon].$$

Proof. By Lemma 1.2 , f has a continuous extension $\tilde{f}\colon \mathbb{R}\times X \to X$ such that $|\tilde{f}(t,x)| \leq M$ everywhere. By Lemma 1.1 , there exists $\tilde{f}_\varepsilon\colon \mathbb{R}\times X \to X$, locally Lipschitz and such that

$$|\tilde{f}_\varepsilon(t,x) - \tilde{f}(t,x)| \leq \varepsilon \qquad ;$$

in particular,

$$|\tilde{f}_\varepsilon(t,x) - f(t,x)| \leq \varepsilon \quad \text{on } J\times D$$

and

$$|\tilde{f}_\varepsilon(t,x)| \leq M + \varepsilon \qquad .$$

Let x_ε be the solution of $x' = \tilde{f}_\varepsilon(t,x)$, $x(0) = x_0$; it exists on $[0,a_\varepsilon]$ and satisfies $x_\varepsilon' = f(t,x_\varepsilon(t)) + y_\varepsilon(t)$ with

$$|y_\varepsilon(t)| = |\tilde{f}_\varepsilon(t,x_\varepsilon(t)) - f(t,x_\varepsilon(t))| \leq \varepsilon \qquad .$$

q.e.d.

This theorem will be used in § 2 and § 3 .

3. Extensions of solutions

In case $\dim X < \infty$ and f is continuous on $J\times X$, it is well known that every solution x of (1) exists either on J or only in some maximal subinterval $[0,\delta_x)$, and then

$$\lim_{t\to\delta_x} |x(t)| = \infty .$$

However, in case dim X = ∞ one can find a continuous f: $[0,\infty)\times X \to X$ such that (1) has a (unique) solution x on $[0,1)$ only, but x remains bounded ; in particular,

$$\lim_{t\to 1} x(t)$$

does not exist. This is possible since a nonlinear continuous mapping need not map bounded sets into bounded sets.

<u>Example 1.1</u>. Let dim X = ∞ . Then there is an infinite dimensional closed subspace $X_o \subset X$ with a Schauder base, i.e. there are sequences $(e_i) \subset X_o$ and $(e_i^*) \subset X_o^*$ (the dual space of X_0) such that each $x \in X_o$ has the unique representation

$$x = \sum_{i\geq 1} \langle x, e_i^* \rangle e_i$$

(see e.g. [108, p.10]) . We may assume $|e_i| = 1$ for every i . We consider the initial value $x_o = e_1 \in X_o$. Then it is enough to construct an f: $[0,\infty)\times X_o \to X_o$ since f has a continuous extension $\tilde{f}$: $[0,\infty)\times X \to X_o$, by Lemma 1.2 , and therefore every solution of (1) with $\tilde{f}$ is already in X_o .

Let $0 < t_1 < t_2 < \ldots < 1$ and

$$\lim_{i\to\infty} t_i = 1 \quad ;$$

χ_i be the characteristic function of $[t_i, t_{i+1}]$; $\phi(t) = \max\{0, 2t-1\}$; $\alpha_1(t) \equiv 1$ and $\alpha_i \in C^1(\mathbb{R})$ for $i \geq 2$ such that

$$\alpha_i(t) = 1 \quad \text{on } [t_i, t_{i+1}] ,$$

$$0 < \alpha_i(t) < 1 \quad \text{on } \left(\frac{t_{i-1}+t_i}{2}, t_i\right) \cup \left(t_{i+1}, \frac{t_{i+1}+t_{i+2}}{2}\right) ,$$

$$\alpha_i(t) = 0 \quad \text{otherwise} .$$

By means of these functions we define

$$f(t,x) = \sum_{i\geq 2} \{\phi(<x,e_{i-1}^*>)\chi_{i-1}(t)+\phi(<x,e_{i+1}^*>)\chi_{i+1}(t)\}\alpha_i'(t)e_i \quad .$$

Since $<x,e_i^*> \to 0$ as $i \to \infty$, it is easy to see that f is continuous and locally Lipschitz on $[0,\infty)\times X_o$. For

$$x(t) = \sum_{i\geq 1} \alpha_i(t)e_i$$

we have $x(0) = e_1$ and

$$x'(t) = \sum_{i\geq 2} \alpha_i'(t)e_i = f(t,x(t)) \quad \text{in } [0,1) \quad .$$

Hence, x is the unique solution of (1) . Since $x(t_i) = e_1 + e_{i-1} + e_i$ and

$$x\left(\frac{t_i+t_{i+1}}{2}\right) = e_1 + e_i \qquad \text{for } i \geq 3 \quad ,$$

$\lim_{t\to 1} x(t)$ does not exist. Finally, $|x(t)| \leq 3$ in $[0,1)$.

4. Linear equations

A simple situation in which the results of section 1 apply is the linear problem

(2) $$x' = A(t)x + b(t) \quad , \quad x(0) = x_o \quad ,$$

in case $A: J \to L(X)$ (the space of bounded linear operators from X into X) and $b: J \to X$ are continuous. Let $R(t,s)$ be the solution of the initial value problem (in $L(X)$) $U' = A(t)U$, $U(s) = I$ (= identity on X) . Then $R(t,s)x_o$ is the solution in J of $x' = A(t)x$, $x(s) = x_o$. By means of this fact it is easy to verify that $R(t,s) = R(t,\tau)R(\tau,s)$ for any $t,\tau,s \in J$, $R(t,s)$ is a homeomorphism of X onto X , $R(t,s)^{-1}=R(s,t)$, and $(t,s) \to R(t,s)$ is continuous. Therefore

$$x(t) = R(t,0)x_o + \int_0^t R(t,s)b(s)ds$$

is the unique solution of (2) . In particular, if $A(t) \equiv A \in L(X)$ then

$$(3) \qquad x(t) = e^{At}x_o + \int_0^t e^{A(t-s)}b(s)ds \quad .$$

This aspect has been veiled sometimes in connection with countable systems of ordinary differential equations. In some papers one did not notice that the corresponding infinite matrix $A = (a_{ij})$ defines a bounded linear operator on the sequence space under consideration, and therefore one had to give a tedious proof e.g. by means of explicite calculation of successive approximations.
For example, let us consider

$$X = l^1 = \{x \in \mathbb{R}^{\mathbb{N}} : \sum_{i \geq 1} |x_i| < \infty\}$$

and suppose that A satisfies

$$\sup_j \sum_{i \geq 1} |a_{ij}| < \infty \quad .$$

Obviously, $A \in L(l^1)$ and therefore the solution of (2) is given by (3).

Now, consider again $X = l^1$, but let us assume that A satisfies

$$\sup_i \sum_{j \geq 1} |a_{ij}| < \infty \quad .$$

Then A is defined on $x \in l^1$ but Ax may not belong to l^1 . Therefore, we can not have a solution of (2) for every $x_o \in l^1$. Nevertheless we may go on at least into the following two directions. On the one hand we may restrict A to its proper domain $D(A) = \{x \in l^1 : Ax \in l^1\}$ and ask wether (2) has a solution at least if x_o and b(t) belong to D(A) . Results of this type will be indicated in § 8 .
On the other hand we may ask wether (2) always has a solution at least in some Banach space larger than that one under consideration ("generalized solutions") . In the present example it is easy to answer this question. Since

$$l^1 \subset l^\infty = \{x \in \mathbb{R}^{\mathbb{N}} : \sup_i |x_i| < \infty\}$$

and $A \in L(l^\infty)$, (2) has a unique solution in l^∞ and (3) is valid. In general, however, the condition that A be bounded from X to some larger Banach space Y such that X is continuously embedded in Y is not sufficient for existence. Consider

Example 1.2. Let $A = (a_{ij})$, where $a_{1j} = 1$ for $j \geq 2$, $a_{i1} = 1$ for $i \geq 2$ and $a_{ij} = 0$ otherwise. Obviously, A is bounded from l^1 into l^∞. If x is a solution of $x' = Ax$, $x(0) = x_o$, then

$$x_i(t) = x_{oi} + \int_0^t x_1(s)ds \qquad \text{for } i \geq 2$$

and

$$x_1'(t) = \sum_{i\geq 2} x_i(t) \quad .$$

Hence, $x_1(t) \equiv 0$ and

$$\sum_{i\geq 2} x_{oi} = 0 \quad .$$

In particular, there is no solution if $x_o \in l^1$ and $x_{o1} \neq 0$.

A positive result in this direction is the following theorem, where one assumes in particular, that A is bounded from the small space to a whole family of larger spaces.

Theorem 1.2. Let (X_s), $0 \leq \alpha \leq s \leq \beta$ be a scale of Banach spaces such that $X_s \subset X_{s'}$ for $s' < s$ and $|x|_{s'} \leq |x|_s$ for $x \in X_s$. Suppose

(i) $A: J \to L(X_s, X_{s'})$ is continuous, for every pair (s,s') with $\alpha \leq s' < s \leq \beta$;

(ii) $|A(t)|_{L(X_s,X_{s'})} \leq \frac{M}{s-s'}$

for some constant $M > 0$ (independent of s,s' and t) ;

(iii) $x_o \in X_\beta$ and $b: J \to X_\beta$ continuous.

Then, for every $s \in [\alpha,\beta)$, (2) has a solution $x: [0,\delta(\beta-s)) \to X_s$, where $\delta = \min\{a, 1/Me\}$. The solution is uniquely determined for $s \in (\alpha,\beta)$, and

$$(*) \qquad |x(t)-x_o|_s \leq (|x_o|_\beta + \frac{\beta-s}{M} \max_{[0,t]} |b(\tau)|_\beta) \frac{Met}{\beta-s-Met} \quad .$$

Proof. 1.) Existence. Consider the successive approximations $x_o(t) \equiv x_o$,

$$x_k(t) = x_o + \int_0^t \{A(\tau)x_{k-1}(\tau) + b(\tau)\}d\tau \quad \text{for } k \geq 1 \quad .$$

By induction, $x_k(t) \in X_s$ for every $s \in [\alpha,\beta)$ and $k \geq 0$. Let

$$M_t = |x_o|_\beta + \frac{\beta-s}{M} \max_{[0,t]} |b(\tau)|_\beta \quad .$$

We claim

$$|x_k(t) - x_{k-1}(t)|_s \leq M_t\left(\frac{tMe}{\beta-s}\right)^k \quad \text{for } k \geq 1 \quad .$$

We have

$$|x_1(t)-x_o|_s \leq t\left(\frac{M}{\beta-s} |x_o|_\beta + \max_{[0,t]} |b(\tau)|_\beta\right) \leq M_t \cdot \frac{Met}{\beta-s} \quad .$$

If the inequality holds for k then

$$|x_{k+1}(t) - x_k(t)|_s \leq \int_0^t |A(\tau)(x_k(\tau) - x_{k-1}(\tau))|_s d\tau$$

$$\leq \frac{M}{\varepsilon} \int_0^t |x_k(\tau) - x_{k-1}(\tau)|_{s+\varepsilon} d\tau \leq M_t\left(\frac{Me}{\beta-s-\varepsilon}\right)^k \frac{M}{\varepsilon} \frac{t^{k+1}}{k+1} \quad ;$$

with $\varepsilon = \frac{\beta-s}{k+1}$ we obtain

$$|x_{k+1}(t) - x_k(t)|_s \leq M_t\left(\frac{Mte}{\beta-s}\right)^{k+1} \cdot e^{-1}\left(1 + \frac{1}{k}\right)^k \leq M_t\left(\frac{Met}{\beta-s}\right)^{k+1} \quad .$$

Hence,

$$x_k(t) \to x(t) := x_o + \sum_{k \geq 1} (x_k(t) - x_{k-1}(t))$$

uniformly on every closed subinterval of $[0,\delta(\beta-s))$, and (*) holds. Therefore,

$$x(t) = x_o + \int_0^t \{A(\tau)x(\tau) + b(\tau)\}d\tau \quad \text{in} \quad [0,\delta(\beta-s)) \quad .$$

Now, $A(t)x(t)$ is continuous, since $x(t)$ is continuous on $[0,\delta(\beta-s-\varepsilon))$ with values in $X_{s+\varepsilon}$ and $t \to A(t) \in L(X_{s+\varepsilon},X_s)$ is continuous, for every $\varepsilon \in (0,\beta-s)$. Hence, x is a solution of (2) .

2.) Uniqueness. Let $s \in (\alpha,\beta)$, and $x: [0,\eta] \to X_s$ satisfy $x' = A(t)x$, $x(0) = 0$. Then $N = \{t : x(t) = 0\}$ is closed. N is also open in $[0,\eta]$. To prove this, let $t_0 \in N$ and $s' < s$. As in the proof of existence we obtain

$$|x(t)|_{s'} \leq M_2\left(\frac{Me|t-t_0|}{s-s'}\right)^k \quad \text{for } k \geq 1 ,$$

by induction, where $M_2 = \max\{|x(t)|_s : t \in [0,\eta]\}$. Hence, $|x(t)|_{s'} = 0$ for $t \in [0,\eta]$ and $|t-t_0| \leq (Me)^{-1}(s-s')$. Since N is open and closed, we have $N = [0,\eta]$.

q.e.d.

We want to illustrate this theorem by a simple example.

Example 1.3. Consider the diffusion equation

$$u_t = u_{xx} + axu_x + bx^2u$$

and the initial condition $u(0,x) = \phi(x)$. If u is interpreted as probability density, for example of a particle undergoing Brownian motion, i.e. $u(t,x) \geq 0$ and

$$\int_{-\infty}^{\infty} u(t,x)dx = 1 ,$$

one is interested in the moments

$$u_n(t) = \int_{-\infty}^{\infty} u(t,x)x^n dx .$$

We multiply the differential equation by x^n and integrate over $\mathbb{R}^1$. Assuming that partial integration is justified, we then obtain the countable system

$$(*) \quad u_n' = n(n-1)u_{n-2} - a(n+1)u_n + bu_{n+2} , \quad u_n(0) = \int_{-\infty}^{\infty} \phi(x)x^n dx ,$$

where $u_{-1}(t) \equiv 0$ and $u_0(t) \equiv 1$. Let A be the corresponding infinite matrix and consider the scale

$$X_s = \{u \in \mathbb{R}^N : |u|_s = \sum_{j\geq 1} |u_j| e^{js}(j!)^{-1/2} < \infty\} \quad \text{for } 0 < \alpha \leq s \leq \beta .$$

Obviously, $|u|_{s'} \leq |u|_s$ for $s' < s$ and $u \in X_s$. A simple calculation yields

$$|A|_{L(X_s,X_{s'})} \leq M(s-s')^{-1} \quad \text{with} \quad M = e^{2\beta}(1+|a|+|b|) \quad ,$$

for instance.
Hence, if the moments of ϕ are in X_β , the moments of u are in X_s for $0 \leq t \leq (Me)^{-1}(\beta-s)$.

5. Excursion to nonlinear eigenvalue problems

In this section we shall indicate how the simple existence theorem for (1) with locally Lipschitz right hand side may be applied to the study of the nonlinear eigenvalue problem $g'(x) = \lambda f'(x)$, where f and g are (Frêchet-) differentiable functionals from a real Hilbert space X into $\mathbb{R}$. A number $\lambda \neq 0$ is called eigenvalue if there exists an $x \neq 0$ such that $g'(x) = \lambda f'(x)$.
Recall first the classical example $g_o(x) = (Tx,x)$ and $f_o(x) = |x|^2$ with a selfadjoint $T \in L(X)$. Here, $g_o'(x) = 2Tx$ and $f_o'(x) = 2x$ and therefore we have the eigenvalue problem for T .
Now, let us consider more general functionals f and g , where f plays the role of f_o and g the role of g_o , and recall a classical theorem on Lagrange multipliers saying that in case g has in x_o a local extremum relative to $\{x \in X : f(x)=f(x_o)\}$ then there exists $\lambda \in \mathbb{R}$ such that $g'(x_o) = \lambda f'(x_o)$, provided $f'(x_o) \neq 0$. If this is the case, we have $\lambda = (g'(x_o),x_o)/(f'(x_o),x_o)$ provided $(f'(x_o),x_o) \neq 0$. Therefore, it is natural to look at

$$(4) \qquad Qx = g'(x) - \frac{(g'(x),x)}{(f'(x),x)} f'(x) \quad \text{on} \quad M_r = \{x \in X : f(x) = r\} \quad ,$$

and $\sup_{M_r} g(x)$. By means of solutions of a certain initial value problem (1) we shall prove the following

Lemma 1.3. Let X be a real Hilbert space ; $f: X \to \mathbb{R}$ and $g: X \to \mathbb{R}$ continuously differentiable and such that

(i) $M_r = \{x \in X : f(x) = r\} \neq \emptyset$ and bounded, for some fixed $r > 0$.
(ii) There are constants $\rho > 0$ and $L > 0$ such that for all $u \in M_r$

$$|f'(x) - f'(y)| \leq L|x-y| \qquad \text{for all } x,y \in \overline{K}_\rho(u) \quad .$$

(iii) For some $\alpha > 0$ we have $(f'(x),x) \geq \alpha$ on $V = \bigcup_{u \in M_r} \overline{K}_\rho(u)$, and $f'(V)$ is bounded.
(iv) $g'(M_r)$ is bounded and there is a continuous increasing function

$d: \mathbb{R}_+ \to \mathbb{R}_+$ with $d(0) = 0$ such that for all $u \in M_r$:

$$|g'(x) - g'(u)| \leq d(|x-u|) \quad \text{for all } x \in \overline{K}_\rho(u) \quad .$$

Then to each $\varepsilon > 0$ there exists $x_\varepsilon \in M_r$ with $g(x_\varepsilon) \geq \sup_{M_r} g(x) - \varepsilon$ such that $|Qx| < \varepsilon$, where Q is defined by (4) .

Once this lemma is proved we have a sequence $(x_n) \subset M_r$ such that $Qx_n \to 0$, and since M_r is bounded we may assume without loss of generality that $x_n \rightharpoonup x_0$ for some $x_0 \in X$, where "$\rightharpoonup$" denotes weak convergence. Now, some standard extra conditions on f and g will yield an eigenvalue $\lambda \neq 0$ with eigenvector x_0 . For example, let us simply assume that g' is strongly continuous, i.e. $v_n \rightharpoonup v$ implies $g(v_n) \to g(v)$, that $|g'(x)| \geq \beta > 0$ on M_r and that f is locally uniformly convex. i.e. to each $x \in X$ there is a continuous function $\phi(x,t)$ with $\phi(x,0) = 0$ and $\phi(x,t) > 0$ for $t > 0$ such that

$$(5) \qquad f(y) - f(x) \geq (f'(x),y-x) + \phi(x,|y-x|) \quad \text{for all } y \in X \; .$$

Then we have $g'(x_n) \to g'(x_0) \neq 0$ and without loss of generality $(g'(x_n),x_n)/(f'(x_n),x_n) \to \lambda$ for some λ . Hence $\lambda f'(x_n) \to g'(x_o) \neq 0$ and therefore $\lambda \neq 0$. Permuting x and y in (5) and adding the result to (5) , we obtain

$$\phi(x_0,|x_n-x_0|) \leq (f'(x_n)-f'(x_0),x_n-x_0) \to 0 \quad \text{as} \quad n \to \infty \quad ,$$

and therefore $x_n \to x_0$. This implies $f'(x_n) \to f'(x_0)$ and therefore $g'(x_0) = \lambda f'(x_0)$.

<u>Proof of Lemma 1.3</u>. Suppose on the contrary that there is an $\varepsilon_0 > 0$ such that $|Qx| \geq \varepsilon_0$ for all $x \in M_r$ with $g(x) \geq c - \varepsilon_0$, where $c = \sup_{M_r} g(v)$. For some $\gamma > 0$ that will be specified later on we choose an $x_0 \in M_r$ with $g(x_0) \geq c - \gamma$ and an $h \in X$ with $|h| = 1$ and $(Qx_0,h) \geq \varepsilon_0/2$. Now, we consider the initial value problem

$$(6) \qquad x' = h - \frac{(f'(x),h)}{(f'(x),x)}\, x =: Fx \quad , \quad x(0) = x_0 \quad .$$

By (ii) and (iii) , it is easy to find constants c_1 and c_2 such that

$$|Fx-Fy| \leq c_1|x-y| \quad \text{and} \quad |Fx| \leq c_2 \quad \text{for all } x,y \in \overline{K}_\rho(x_0) \; .$$

Hence, (6) has a unique solution $x(t)$ in $0 \leq t \leq t_0 = \rho/c_2$. Furthermore, $x(t) \in M_r$ in $[0,t_o]$ since $[f(x(t))]' = (f'(x(t)),x'(t)) = 0$ and $f(x(0)) = r$. With $y(s) = x_o + s(x(t)-x_o)$ we have

$$g(x(t)) - g(x_o) = \int_0^1 (g'(y(s)),x(t)-x_o)ds = \int_0^t \left(\int_0^1 g'(y(s))ds, x'(\tau)\right) d\tau$$

$$\geq \int_0^t (g'(x_o),x'(\tau))d\tau - c_2 t \sup_{s\in[0,1]} |g'(y(s))-g'(x_o)| \ .$$

Now,

$$(g'(x_o),x'(\tau)) = (g'(x_o),h) - \frac{(f'(x),h)}{(f'(x),x)}(g'(x_o),x)$$

$$\geq (Qx_o,h) - \left| \frac{(g'(x_o),x_o)}{(f'(x_o),x_o)}(f'(x_o),h) - \frac{(f'(x),h)}{(f'(x),x)}(g'(x_o),x) \right|$$

for $x = x(\tau)$.

A simple calculation yields $|\ldots| \leq c_3|x(\tau)-x_o| \leq c_3c_2\tau$ for some constant c_3 . Hence

$$g(x(t)) \geq c - \gamma + \frac{\varepsilon_o}{2} t - c_3 \frac{c_2}{2} t^2 - c_2 t d(c_2 t) \ .$$

We notice that by (ii)-(iv) the constants c_i are independent of $x_o \in M_r$. Therefore, we can choose t_1 so small that

$$\delta(t_1) = t_1\left(\frac{\varepsilon_o}{2} - c_3\frac{c_2}{2} t_1 - c_2 d(c_2 t_1)\right) > 0$$

and now $\gamma > 0$ such that $\gamma < \delta(t_1)$. Then we have $g(x(t_1)) > c$, a contradiction.

q.e.d.

6. <u>Remarks</u>

(i) Lemma 1.1 is taken from Lasota/Yorke [101] . Lemma 1.2 is due to Dugundji [57] and also proved in Deimling [46 ,p.21] .

(ii) The phenomenon of Example 1.1 has been observed by Dieudonné [55] who has an example for $X = (c_o)$, the space of all sequences tending to zero. This phenomenon is also known for functional differential equations ; see Yorke [193] for an elegant construction of examples. For the l^1-example in sect. 4 consider Shaw [159] . Example 1.2 is taken from Arley/Borchsenius [4] .

(iii) Theorem 1.2 is due to Ovcyannikov [135] . A detailed discussion of this theorem may be found in Treves [171] . Obviously, the theorem may also be stated for complex t . Then J is a disc in $\mathbb{C}$ with center $t = 0$, A(t) and b(t) are supposed to be holomorphic and the corresponding solutions x_s are holomorphic in the discs with radius $\delta(\beta-s)$. Applications to moment problems like that in Example 1.3 are given in Steinberg/Treves [165] and Steinberg [164] . Obviously, we may replace the linear right hand side $A(t)x + b(t)$ of Theorem 1.2 by a nonlinear f such that $f: J\times X_s \to X_{s'}$ is continuous and

$$|f(t,x)-f(t,y)|_{s'} \leq \frac{M}{(s-s')}|x-y|_s \quad \text{for} \quad x,y \in X_s \quad \text{and} \quad t\in J \quad .$$

Further generalizations and applications are given by DuChateau [58] . In § 2 we shall consider some nonlinear perturbations of the linear problem.

(iv) Lemma 1.3 is a special case of Naumann [129,Lemma 3] . To understand better what the trajectory of (6) was good for let us sketch a little bit more "Lusternik- Shnirelman Theory" for the eigenvalue problem. Let us start with g_o, f_o of section 5 and assume in addition that T is compact and $(Tx,x) > 0$. Then the classical max-min-principle tells us that the eigenvalues λ_n of T are given by

$$\lambda_n = \max_{U\in \mathfrak{U}_n} \min_{x\in U\setminus\{0\}} \frac{(Tx,x)}{(x,x)} = \frac{1}{r^2} \max_{U\in \mathfrak{U}_n} \min_{x\in U\cap S_r} g_o(x) \quad (n = 1,2,\dots) \quad ,$$

where $\mathfrak{U}_n = \{U : U \text{ is subspace of } X \text{ with } \dim U = n\}$ and $S_r = \partial\overline{K}_r(0)$; see e.g. [190,chap.3] , [73,p.304] . In order to simulate this technique for more general f and g , we need a certain class $\mathcal{V}$ of subsets of M_r which plays the role of $\{U\cap S_r : U\in\mathfrak{U}_n\}$. Several people have used a deformation invariant class $\mathcal{V}$, i.e. an $\mathcal{V}$ such that if $S\in\mathcal{V}$ and $H: M_r\times[0,1] \to M_r$ is continuous and $H(\cdot,0) = I|_{M_r}$ then $H(S,1)\in\mathcal{V}$. A simple example is $\mathcal{V}_o = \{\{x\}: x\in M_r\}$. Then

$$c = \sup_{S\in\mathcal{V}_o} \inf_{x\in S} g(x) = \sup_{M_r} g(x) \qquad .$$

Another example is

$$\mathcal{V} = \{S\subset M_r : \text{cat}(S,M_r) \geq k\}$$

for some integer $k \geq 1$, where $\text{cat}(S,M_r)$ denotes the Lusternik-Shnirelman category of S in M_r ; see e.g. Browder [24] , Rabinowitz [145] . In the proof of Lemma 1.3 we have, generally spoken, used $\mathcal{V}_o$, $S_o=\{x_o\}\in\mathcal{V}_o$ such that $\inf_{S_o} g(x) \geq c-\gamma$, and a deformation H of S_o to $S_1 = \{x(t_1)\}$

with $\inf_{S_1} g(x) > c$, the deformation $H: M_r \times [0,1] \to M_r$ being defined by

$$H(x_o, s) = x(st_1; x_o) \quad ,$$

where $x(\cdot; x_o)$ denotes the solution of (6) . In this way Naumann proved Lemma 1.3 for any deformation invariant class $\mathcal{V}$. Further results on the eigenvalue problem may be found e.g. in the lecture notes [67] .

§ 2 Compactness Conditions

1. Nonexistence.

Let X be a Banach space, $D = \overline{K}_r(x_o) \subset X$, $J = [0,a] \subset \mathbb{R}$ and $f: J \times D \to X$ continuous. If $\dim X < \infty$ then the initial value problem

$$(1) \qquad x' = f(t,x) \quad , \quad x(0) = x_o$$

has a local solution. All proofs of this well known 'Peano'-Theorem essentially depend on a compactness argument, the Ascoli/Arzelà-Theorem which we state as

Lemma 2.1. Let X be a Banach space and $C_X(J)$ the space of all continuous functions $x: J \to X$ with norm $|x|_o = \max_J |x(t)|$. A subset $M \subset C_X(J)$ is relatively compact if and only if M is equicontinuous and

$$M(t) = \{x(t) : x \in M\}$$

is relatively compact in X for every $t \in J$.

Consider, for instance, the approximate solutions x_ε in Theorem 1.1, where f is bounded on $J \times D$. Obviously, $\{x_\varepsilon : 0 < \varepsilon \leq 1\} \subset C_X([0,b])$ is equicontinuous and bounded. Hence, Lemma 2.1 applies in case $\dim X < \infty$. If, however, $\dim X = \infty$ then a bounded subset of X need not be relatively compact. This indicates that the Peano-Theorem may be wrong for $\dim X = \infty$. In fact, consider the following simple

Example 2.1. Let $X = (c_o)$ the space of all real sequences with limit zero and $|x| = \max_j |x_j|$, let $e_i \in X$ be defined by $e_{ij} = \delta_{ij}$ (Kronecker symbol). Then

$$x = (x_j) = \sum_{j \geq 1} x_j e_j$$

for each $x \in X$. Let

$$x_o = \sum_{i \geq 1} i^{-2} e_i \qquad \text{and} \qquad f(x) = 2 \sum_{i \geq 1} \sqrt{|x_i|}\, e_i \, .$$

Suppose x is a local solution of (1). Then

$$x_i' = 2\sqrt{|x_i|} \quad \text{and} \quad x_i(0) = i^{-2} \quad .$$

Hence, $x_i(t) = (t + \frac{1}{i})^2$ for every i and $t \geq 0$, in particular

$$\lim_{i\to\infty} x_i(t) = t^2 \neq 0 \quad \text{for} \quad t > 0 \quad ,$$

a contradiction.

However, if we assume that f maps $J\times D$ into a relatively compact set and (ε_n) is any sequence with $\varepsilon_n > 0$ and $\varepsilon_n \to 0$ then $(x_{\varepsilon_n}(t))$ is relatively compact in X for $t \leq b$, and we find a solution of (1) as in the finite dimensional case. Since loss of compactness is the main reason for nonexistence, it is natural to 'measure' the noncompactness of subsets of X and to look for conditions on f , sufficient for existence, in terms of such a measure.

2. Measures of noncompactness

Let us recall that a subset M of a Banach space X is relatively compact iff to every $\varepsilon > 0$ there are finitely many balls of radius ε such that their union covers M . If M is only bounded, there is a positive lower bound for such numbers ε . These facts suggest the following

Definition 2.1. Let X be a Banach space and $\mathcal{B}$ the family of all bounded subsets of X . Then the (Hausdorff-) measure of noncompactness $\gamma: \mathcal{B} \to \mathbb{R}$ is defined by

$$\gamma(B) = \inf\{\varepsilon > 0 : B \text{ admits a finite } \varepsilon\text{-ball covering}\} \quad .$$

The (Kuratowski-) measure of noncompactness $\alpha: \mathcal{B} \to \mathbb{R}$ is defined by

$$\alpha(B) = \inf\{d > 0 : B \text{ admits a finite covering by sets of diameter } \leq d\} \quad .$$

These measures of noncompactness have several properties which will be useful in the sequel.

Lemma 2.2. Let $\alpha: \mathcal{B} \to \mathbb{R}$ be as in Def. 2.1 . Then

(i) $\alpha(B) = 0 \iff \overline{B}$ is compact ; $\alpha(\overline{K}_1(0)) \leq 2$.

(ii) α is a seminorm, i.e. $\alpha(\lambda B) = |\lambda|\alpha(B)$ and $\alpha(B_1+B_2) \leq \alpha(B_1) + \alpha(B_2)$.

(iii) $B_1 \subset B_2 \Rightarrow \alpha(B_1) \leq \alpha(B_2)$; $\alpha(B_1 \cup B_2) = \max\{\alpha(B_1), \alpha(B_2)\}$.
(iv) $\alpha(\text{conv } B) = \alpha(B)$.
(v) α is continuous with respect to the Hausdorff metric

$$d_H(B_1,B_2) = \max\{\sup_{B_1} \rho(x,B_2), \sup_{B_2} \rho(x,B_1)\} \quad .$$

In particular, $\alpha(B) = \alpha(\overline{B})$.

The measure γ has the same properties.

Proof. (i) - (iii) and (v) are immediate consequences of Def. 2.1 . To prove (iv) , we only have to show $\alpha(\text{conv } B) \leq \alpha(B)$, by (iii) . Let $d > \alpha(B)$ and

$$B \subset \bigcup_{i=1}^{m} M_i \quad \text{with} \quad \text{diam } M_i \leq d \quad .$$

Since $\text{diam}(\text{conv } M_i) \leq d$, we may assume that M_i is convex. Since

$$\text{conv } B \subset \text{conv}[M_1 \cup \text{conv}(\bigcup_{i=2}^{m} M_i)] \subset \text{conv}[M_1 \cup \text{conv}[M_2 \cup \text{conv}(\bigcup_{i=3}^{m} M_i)]] \subset \ldots$$

it is sufficient to show $\alpha(\text{conv}(C_1 \cup C_2)) \leq \max\{\alpha(C_1), \alpha(C_2)\}$ for convex C_1 , C_2 . It is easy to see that

$$\text{conv}(C_1 \cup C_2) = \bigcup_{\lambda \in [0,1]} (\lambda C_1 + (1-\lambda)C_2) \quad .$$

Since $C_1 - C_2$ is bounded, we have $|x| \leq k$ for some $k > 0$ and every $x \in C_1 - C_2$. Given $\varepsilon > 0$, we find $\lambda_1, \ldots, \lambda_p$ such that

$$[0,1] \subset \bigcup_{i=1}^{p} K_{\varepsilon/k}(\lambda_i) \quad .$$

This implies

$$\text{conv}(C_1 \cup C_2) \subset \bigcup_{i=1}^{p} [\lambda_i C_1 + (1-\lambda_i)C_2 + \overline{K}_\varepsilon(0)] \quad .$$

Hence, by (iii) and (ii) ,

$$\alpha(\text{conv}(C_1 \cup C_2)) \leq \max\{\alpha(C_1), \alpha(C_2)\} + 2\varepsilon \quad .$$

q.e.d.

3. Existence

Suppose $f: J\times X \to X$ is continuous and maps bounded sets into relatively compact sets. We call such an f '<u>compact</u>'. Let $f(t,B) = \{f(t,x): x\in B\}$. Then $\alpha(f(t,B)) = 0 \leq \alpha(B)$ for $B\in \mathcal{B}$. Now, suppose f admits a splitting $f = f_1+f_2$ such that f_1 is compact and f_2 is Lipschitz ,

$$|f_2(t,x) - f_2(t,y)| \leq k|x-y| \quad .$$

Then, by Lemma 2.2 ,

$$\alpha(f(t,B)) \leq \alpha(f_1(t,B)+f_2(t,B)) \leq \alpha(f_2(t,B)) \leq k\alpha(B) \quad \text{for} \quad B\in\mathcal{B} .$$

We shall show now that estimates of this kind imply local existence for problem (1) . To this end we start with

<u>Proposition 2.1</u>. Let $x: [0,a] \to X$ be differentiable. Then

$$\{\frac{x(t)-x(t-h)}{h} : t\in(0,a] \ , \ 0 < h \leq t\} \subset \overline{\text{conv}}\{x'(t) : t\in[0,a]\} \quad .$$

<u>Proof</u>. Let K be the set on the right hand side. It is closed and convex, and therefore the intersection of all half spaces

$$\{x\in X : \text{Re } x^*(x) \leq \lambda\}$$

which contain K . Hence, we have to show

$$\text{Re } x^*(x(t)-x(t-h)) \leq \lambda h \quad \text{whenever } K\subset\{x : \text{Re } x^*(x) \leq \lambda\} \quad .$$

Let $t_0\in[0,a)$ and

$$\phi(t) = \text{Re } x^*(x(t)-x(t_0)) \quad \text{for} \quad t\in[t_0,a] \quad .$$

We have $\phi'(t) = \text{Re } x^*(x'(t)) \leq \lambda$ and $\phi(t_0) = 0$, hence $\phi(t) \leq \lambda(t-t_0)$.

q.e.d.

<u>Theorem 2.1</u>. Let X be a Banach space, $D = \overline{K}_r(x_0)\subset X$, $J = [0,a]\subset \mathbb{R}$, $f: J\times D \to X$ uniformly continuous and bounded, say $|f(t,x)| \leq c$. Suppose there is a constant $L \geq 0$ such that

$$(2) \qquad \alpha(f(t,B)) \leq L\alpha(B) \quad \text{for all} \quad t\in J \ , \quad B\subset D \ ,$$

and let $b < \min\{a,r/c\}$. Then, problem (1) has a solution on $[0,b]$.

<u>Proof</u>. By Theorem 1.1 , we have approximate solutions x_n on $[0,b]$ such that

$$x_n' = f(t,x_n(t))+y_n(t) \ , \ x_n(0) = x_0 \text{ and } |y_n(t)| \leq 1/n \quad .$$

Obviously, the sequence (x_n) is equicontinuous. Hence, by Lemma 2.1, we have to show that $\{x_n(t) : n \geq 1\}$ is relatively compact, i.e. $\alpha(\{x_n(t) : n \geq 1\}) = 0$ for every $t \in [0,b]$. Then, a standard argument shows that the limit of the existing uniformly convergent subsequence is a solution of (1) on $[0,b]$. Let

$$B_k(t) = \{x_n(t) : n \geq k\} \quad , \quad B_k'(t) = \{x_n'(t) : n \geq k\}$$

and

$$\phi_k(t) = \alpha(B_k(t)) \quad .$$

We have $\phi_k(0) = 0$ and ϕ_k continuous, since, by Lemma 2.2 (ii)

$$|\phi_k(t)-\phi_k(\bar{t})| \leq \alpha(\{x_n(t)-x_n(\bar{t}) : n \geq k\}) \leq 2(c+1)|t-\bar{t}| \quad .$$

We claim

$$D^-\phi_k(t) \leq \alpha(f(t,B_k(t))) + 2/k \quad \text{in} \quad (0,b] \quad ,$$

where

$$D^-\phi_k(t) = \limsup_{\tau\to 0+ (0,\tau]} q(h) \quad \text{with} \quad q(h) = h^{-1}(\phi_k(t)-\phi_k(t-h)) \quad .$$

By Lemma 2.2 (ii) and (iii), we obtain

$$\sup_{(0,\tau]} q(h) \leq \alpha\left(\left\{\frac{x_n(t)-x_n(t-h)}{h} : n \geq k \; , \; 0 < h \leq \tau\right\}\right) \quad .$$

By Proposition 2.1 and property (iv) of $\alpha(\cdot)$, this estimate implies

$$D^-\phi_k(t) \leq \lim_{\tau\to 0+} \alpha\Big(\bigcup_{J_\tau} B_k'(s)\Big) \quad \text{with} \quad J_\tau = [t-\tau,t] \quad .$$

Evidently,

$$(*) \qquad \begin{aligned} \alpha\Big(\bigcup_{J_\tau} B_k'(s)\Big) &\leq \alpha\Big(\bigcup_{J_\tau} f(s,B_k(s)) + \bigcup_{J_\tau}\bigcup_{n\geq k} \{y_n(s)\}\Big) \leq \\ &\leq \alpha\Big(\bigcup_{J_\tau} f(s,B_k(s))\Big) + 2/k \quad . \end{aligned}$$

Since (x_n) is equicontinuous and f is uniformly continuous,

$$\bigcup_{J_\tau} f(s,B_k(s)) \to f(t,B_k(t))$$

with respect to the Hausdorff metric. Therefore

$$D^-\phi_k(t) \leq \alpha(f(t,B_k(t)) + 2/k \quad .$$

By (2), we have $\alpha(f(t,B_k(t))) \leq L\phi_k(t)$. Hence, $\phi_k(t) \leq 2(kL)^{-1}e^{Lt}$, and therefore

$$\alpha(\{x_n(t) : n \geq 1\}) = \alpha(B_k(t)) \leq 2(kL)^{-1}e^{Lt} \to 0 \quad \text{as} \quad k \to \infty \; .$$

q.e.d.

Notice that a continuous $f\colon J\times\overline{K}_r(x_o) \to X$ is bounded if r is sufficiently small. But f need not be uniformly continuous if $\dim X = \infty$. It is easy to see that the uniform continuity of f implies

$$\alpha(f(J\times B)) = \max_J \alpha(f(t,B)) \quad \text{for} \quad B \subset D \; .$$

Hence, Theorem 2.1 is a particular case of our next theorem, where we dispense with uniform continuity. We have given a seperate proof since it is valid for much more general right hand sides in (2) ; see Remark (iii) at the end.

Theorem 2.2. Let X,D,J,c and b be as in Theorem 2.1 . Suppose $f\colon J\times D \to X$ is continuous and satisfies

(3) $$\alpha(f(J\times B)) \leq \omega(\alpha(B)) \qquad \text{for} \quad B \subset D \; ,$$

where $\omega\colon \mathbb{R}_+ \to \mathbb{R}_+$ is continuous and such that the initial value problem $\rho' = \omega(\rho)$, $\rho(0) = 0$ has only the trivial solution $\rho(t) \equiv 0$ on J . Then, (1) has a solution on $[0,b]$.

Proof. We follow the proof of Theorem 2.1 up to (*) . By means of (3) we may continue

$$\alpha\Big(\bigcup_{J_\tau} f(s,B_k(s))\Big) \leq \alpha\Big(f([0,b]\times \bigcup_{J_\tau} B_k(s))\Big) \leq \omega\Big(\alpha\Big(\bigcup_{J_\tau} B_k(s)\Big)\Big) \; .$$

Hence, $D^-\phi_k(t) \leq \omega(\phi_k(t)) + 2/k$ in $(0,b]$ and $\phi_k(0) = 0$. This implies $\phi_k(t) \to 0$ as $k \to \infty$, since, by our assumption on ω , there are functions ρ_n such that

$$\rho_n' = \omega(\rho_n) + \frac{1}{n} \quad , \quad \rho_n(0) = \frac{1}{n}$$

and $\rho_n(t) \to 0$ as $n \to \infty$, and $\phi_k(t) < \rho_n(t)$ for $k > 2n$.

q.e.d.

4. The set of solutions

The conditions of Theorem 2.1 and Theorem 2.2 do not ensure uniqueness. Therefore, it is interesting to look for properties of S , the set of all solutions of problem (1) . We shall show that S is a compact and connected subset of $C_X([0,b])$. This result is the extension to higher dimensions of the well known one dimensional Peano funnel.

Theorem 2.3. Let the conditions of Theorem 2.1 or of Theorem 2.2 be satisfied. Then, the set $S \subset C_X([0,b])$ of all solutions on $[0,b]$ of (1) is a continuum. In particular, $S(t) = \{x(t) : x \in S\}$ is a continuum of X , for every $t \in [0,b]$.

Proof. If we consider a sequence of solutions of (1) , instead of the approximate solutions x_n in the existence proof, we obtain a uniformly convergent subsequence, and its limit is again a solution. Hence, S is compact.
Suppose S is not connected. Then $S = S_1 \cup S_2$ with $S_1 \cap S_2 = \emptyset$ and S_1 , S_2 compact. Hence, $\beta = \rho(S_1,S_2) > 0$. Consider the functional

$$\phi: C_X([0,b]) \to \mathbb{R} \quad , \text{defined by} \quad \phi(x) = \rho(x,S_1) - \rho(x,S_2) \quad .$$

Obviously, ϕ is continuous and $\phi(x) \leq -\beta$ on S_1 , $\phi(x) \geq \beta$ on S_2 . We are going to construct an $x \in S$ such that $\phi(x) = 0$, a contradiction. Therefore, S is connected. Since S is a continuum and $x \to x(t)$ is continuous, $S(t)$ is a continuum in X .
To construct $x \in S$ with $\phi(x) = 0$, let $\varepsilon > 0$ be such that $(c+2\varepsilon)b \leq r$. By Lemma 1.1 , we find $g_\varepsilon: J\times D \to X$ locally Lipschitz and such that $|g_\varepsilon(t,x) - f(t,x)| \leq \varepsilon$ on $J\times D$. Let $x_i \in S_i$ be fixed and consider

$$f^i(t,x) = g_\varepsilon(t,x) + f(t,x_i(t)) - g_\varepsilon(t,x_i(t)) \quad \text{for} \quad i = 1,2 \; .$$

For $\lambda \in [0,1]$, let

$$f_\lambda(t,x) \;=\; f^1(t,x) + \lambda(f^2(t,x) - f^1(t,x)) \quad .$$

We have f_λ locally Lipschitz and $|f_\lambda(t,x) - f(t,x)| \leq 2\varepsilon$. Therefore, the problem $x' = f_\lambda(t,x)$, $x(0) = x_o$ has a unique solution x^λ on $[0,b]$, and it is easy to see that $\lambda \to x^\lambda$ is continuous. Hence, $\psi(\lambda) = \phi(x^\lambda)$ is continuous in $[0,1]$. Since

$$f_o(t,x_1(t)) = f^1(t,x_1(t)) = x_1'(t) \quad ,$$

we have $x^o = x_1$, by uniqueness. Similarly, $x^1 = x_2$. Hence, $\psi(0) \leq -\beta$ and $\psi(1) \geq \beta$. Therefore, we find $\lambda(\varepsilon) \in (0,1)$ such that $\psi(\lambda(\varepsilon)) = 0$.
Let

$$x_n = x^{\lambda(\varepsilon_n)} \quad \text{for some sequence } \varepsilon_n \to 0+ \quad .$$

Since

$$x_n' = f(t,x_n(t)) + y_n(t) \quad , \quad x_n(0) \;=\; x_o$$

and

$$|y_n(t)| = |f_{\lambda(\varepsilon_n)}(t,x_n(t)) - f(t,x_n(t))| \leq 2\varepsilon_n \quad ,$$

(x_n) has a subsequence, uniformly convergent to a solution x of (1) , i.e. $x \in S$ and $\phi(x) = \lim_{n\to\infty} \phi(x_n) = 0$.

q.e.d.

5. Excursion to Ovcyannikov

Evidently, Theorem 2.1 does not change if we consider $A(t)x + f(t,x)$ instead of f , with $A: J \to L(X)$ continuous. It is possible to generalize this result to the situation given by Ovcyannikov's theorem. To prove such a perturbation theorem we shall need the following extension of Schauder's fixed point theorem

Lemma 2.3. Let X be a Banach space, $K \subset X$ bounded closed and convex, $T: K \to K$ continuous and such that for some constant $k \in [0,1)$

$$\alpha(T(B)) \leq k\alpha(B) \qquad \text{for all} \quad B \subset K \quad .$$

Then T has a fixed point.

It is easy to prove this Lemma by means of Schauder's fixed point theorem and the following simple

Proposition 2.2. Let X be a Banach space and (A_n) a sequence of closed subsets of X such that $A_1 \supset A_2 \supset \ldots$ and $\alpha(A_n) \to 0$ as $n \to \infty$. Then

$$A = \bigcap_{n\geq 1} A_n \neq \emptyset \quad \text{and A is compact.}$$

Proof. Pick $x_n \in A_n$ for each $n \geq 1$. Since (A_n) is decreasing, we have

$$\alpha(\{x_n : n \geq 1\}) = \alpha(\{x_n : n \geq k\}) \leq \alpha(A_k) \to 0 \quad \text{as} \quad k \to \infty \quad .$$

Hence, $\{x_n : n \geq 1\}$ is relatively compact and therefore, without loss of generality, $x_n \to x$ for some $x \in X$. Since the A_k are closed, we have $x \in A_k$ for every k , and therefore $x \in A$. Finally, $\alpha(A) \leq \alpha(A_k) \to 0$. Hence, A is compact.

q.e.d.

Proof of Lemma 2.3. Let $A_1 = \overline{\text{conv}\, T(K)}$ and $A_n = \overline{\text{conv}\, T(A_{n-1})}$ for $n \geq 2$. Evidently, (A_n) is a decreasing sequence of closed convex subsets, and therefore $C = \bigcap_{n\geq 1} A_n$ is closed convex. Since

$$\alpha(A_n) = \alpha(T(A_{n-1})) \leq k\alpha(A_{n-1}) \leq \ldots \leq k^n\alpha(K) \to 0 \quad \text{as} \quad n \to \infty \quad ,$$

C is also compact, by Proposition 2.2 . Moreover, $T: C \to C$, and T is continuous. Hence, by Schauder's theorem T has a fixed point $x \in C \subset K$.

q.e.d.

Now, we are in position to prove

Theorem 2.4. Let the scale $(X_s)_{\mu \leq s \leq \nu}$ of Banach spaces and the linear operators A(t) be as in Theorem 1.2 (in particular: X_ν is the smallest space) ;

$$D = \overline{K}_r(x_o) \subset X_s \quad \text{for some fixed } s \in (\mu,\nu) \quad ;$$

$J = [0,a]$ and $f: J \times D \to X_\nu$ uniformly continuous ;

$$|f(t,x)|_\nu \leq c \qquad \text{on} \quad J \times D$$

and

(4) $$\alpha_\nu(f(t \times B)) \leq L\alpha_s(B) \qquad \text{for} \quad t \in J \ , \ B \subset D \quad ,$$

where $\alpha_s(\cdot)$ denotes the measure of noncompactness for X_s . Then,

$$x' = A(t)x + f(t,x) \quad , \quad x(0) = x_o$$

has a local solution in X_s .

Proof. Let $E = C_{X_s}(J)$. Consider $u \in E$ with range in D . By Theorem 1.2,

(5) $$x' = A(t)x + f(t,u(t)) \quad , \quad x(0) = x_o$$

has a unique solution on $[0,\delta_o(\nu-\tau))$ with range in X_τ , for every $\tau \in (\mu,\nu)$, where $\delta_o = \min\{a,(Me)^{-1}\}$.

We choose $\varepsilon > 0$ and $\delta > 0$ such that $s+\varepsilon < \nu$, $\delta < \delta_o(\nu-s-\varepsilon)$ and $\delta \leq a$. For functions $u \in E$ with range in D we define a mapping T by

$$(Tu)(t) = \begin{cases} \text{the solution of (5)} & \text{for } t \in [0,\delta] \\ (Tu)(\delta) & \text{for } t \in [\delta,a] \end{cases} \quad .$$

Let

$$c_1 = \frac{Me\delta}{(\nu-s-\varepsilon-Me\delta)} \quad .$$

By Theorem 1.2 , we have

(6) $$|(Tu)(t)-x_o|_s \leq c_2c_1 \quad , \quad \text{where } c_2 = |x_o|_\nu+(\nu-s)M^{-1}c$$

(7) $$|(Tu)(t)-(Tv)(t)|_s \leq c_1(\nu-s)M^{-1} \max_{[0,\delta]}|f(\tau,u(\tau))-f(\tau,v(\tau))|_\nu .$$

Since (Tu)(t) is also in the smaller space $X_{s+\varepsilon}$, we can apply (6) for

$s+\varepsilon$ to obtain by means of the integral equation for Tu

$$(8) \qquad |(Tu)(t)-(Tu)(\bar{t})|_s \leq c_3|t-\bar{t}| \quad , \text{ where } c_3 = M\varepsilon^{-1}[|x_0|_\beta + c_2 c_1] + c \ .$$

The estimates (6) - (8) are valid for $t,\bar{t} \in J$. Now, we choose δ also so small that $c_2 c_1 \leq r$ and $c_1(\nu-s)M^{-1}L < 1$, where L is from (4) . Let

$$K = \{u \in E : \max_J |u(t)-x_0|_s \leq r \ , \ |u(t)-u(\bar{t})|_s \leq c_3|t-\bar{t}| \text{ for } t,\bar{t} \in J\}.$$

The set K is closed bounded and convex. By (6) - (8) , T is a continuous mapping from K into K . Consider the map F , defined by

$$(Fu)(t) = f(t,u(t)) \ .$$

Let $B \subset K$ and α_E the Kuratowski-measure for E . By (7) , we have

$$\alpha_E(T(B)) \leq c_1(\nu-s)M^{-1}\alpha_{C_{X_\nu}(J)}(F(B)) \ .$$

Since f is uniformly continuous and B is equicontinuous, F(B) is equicontinuous too. This implies

$$\alpha_{C_{X_\nu}(J)}(F(B)) = \max_J \alpha_\nu(F(B)(t)) \ .$$

By (4) , $\alpha_\nu(F(B)(t)) = \alpha_\nu(f(t,B(t)) \leq L\alpha_s(B(t))$. Hence, we have

$$\alpha_E(T(B)) \leq k\alpha_E(B) \quad , \quad \text{where } k = c_1(\nu-s)M^{-1}L < 1$$

by our choice of δ . Therefore, we may apply the fixed point Lemma 2.3: T has a fixed point, and this fixed point is a solution on $[0,\delta]$ of the initial value problem.

q.e.d.

6. Remarks

(i) Example 2.1 is due to Dieudonné [55] . Yorke [196] has an example for l^2 which was slightly simplified in Lasota/Yorke [101] , considering $L^2(0,\infty)$ instead of l^2 . Cellina [29] has given an elegant construction of counterexamples in nonreflexive spaces X . If X is not reflexive then there exists $x^* \in X^*$ such that $|x^*| = 1$ and $x^*(x) < 1$ for every $x \in D = \bar{K}_1(0)$; see James [80] . By means of x^* , Cellina defines a continuous map $g: D \to D$ without fixed points and a continuous extension $\tilde{g}: X \to D$ (by Lemma 1.2) . Then, problem (1) with

$$f(t,x) = 2t\tilde{g}(x/t^2) \quad \text{and} \quad x_0 = 0$$

has no local solution. Recently, Godunov [68] published the following counterexample for general X :

As in Example 1.1 we may assume that X has a Schauder base (e_i, e_i^*) with $|e_i| = 1$ for every $i \geq 1$. Let $x_0 = 0$ and

$$f(t,x) = \begin{cases} |P(t,x)|^{-1/2}P(t,x) + \phi(t,x) & \text{for } P(t,x) \neq 0 \\ \phi(t,x) & \text{for } P(t,x) = 0 \end{cases} ;$$

$$P(t,x) = \sum_{n\geq 1} \psi_n(t)<x,e_n^*>e_n \quad ;$$

$\psi_n(t) = 0$ for $t \leq c_n$, $= 1$ for $t \geq b_n$ and linear in $[c_n,b_n]$, where

$$a_n = \frac{1}{2n+1} \quad , \quad b_n = \frac{1}{2n} \quad \text{and} \quad c_n = \frac{a_n+b_n}{2} \quad .$$

The function ϕ is defined by

$$\phi(t,x) = \sum_{n\geq 1} \phi_n(t)\psi\left(\frac{(t-b_{n+1})^2}{4} - |x-P_nx|\right)e_n \quad ,$$

where

$$P_nx = \sum_{i=1}^{n} <x,e_i^*>e_i \quad \text{and} \quad \psi(t) = \begin{cases} 0 & \text{for } t < 0 \\ t & \text{for } 0 \leq t \leq 1 \\ 1 & \text{for } t > 1 \end{cases} ,$$

ϕ_n is continuous and such that $\phi_n(t) = 0$ for $t \notin (a_n,c_n)$, $0 < \phi_n(t) < 1/n$ in (a_n,c_n) .

Then problem (1) can not have a local solution since "$x(b_n) \neq 0$ for infinitely many b_n" is impossible as well as "$x(b_n) = 0$ for all sufficiently large n" .

(ii) By means of Lemma 1.1 , Lasota/Yorke [101] have shown that existence of solutions for problem (1) is at least a generic property in the following sense.

Let X be a Banach space ; $\Omega \subset \mathbb{R}\times X$ open ; E the space of continuous $f: \Omega \to X$ with the topology of uniform convergence. Let $B \subset \Omega$ be a countable union of compact sets. Then, there exists a subset $E_1 \subset E$ such that $E \setminus E_1$ is meager and the following is true: For $f \in E_1$ and $x_0 \in B$ the initial value problem has an unlimited solution which is unique and depends continuously on f and x_0 . A solution x of $x' = f(t,x)$, $x(t_0) = x_0$, existing in some interval (a,b) with $-\infty \leq a < b \leq \beta$, is called unlimited if it has no limits in Ω as $t \to a$ and $t \to b$.

They have also proved that the set of functions $f \in E$, satisfying condition (2) for some t , is meager in case $\dim X = \infty$. For more results in this direction we refer to Vidossich [176] and De Blasi/Myjak [45] .

(iii) For more details about measures of noncompactness we refer to the survey article of Sadovski [158] . The fixed point Lemma 2.3 has been proved earlier by Darbo [42] , while Proposition 2.2 may already be found in Kuratowski [93] and in Nußbaum [132] , for example.
Continuous bounded mappings $f: X \to X$ such that $\alpha(f(B)) \leq k\alpha(B)$ for some constant k and all $B \in \mathcal{B}$ are usually called k-set contractions. As a simple example we have already mentioned $f = f_1+f_2$, where f_2 is compact and f_1 is Lipschitz with constant k . More general are mappings of the following type. Let $V: X\times X \to X$ be such that
(i) $V(\cdot,y)$ is Lipschitz with a fixed constant k for each $y \in X$
(ii) $V(x,\cdot)$ is compact for each $x \in X$.
Then it is easy to see that $f: X \to X$, defined by $f(x) = V(x,x)$, is a 2k-set contraction. See Browder [21, Chap.13] and Nußbaum [132] for more examples of this type.

(iv) The proof of Theorem 2.1 is such that it works for more general estimates (2) . Consider

$$(9) \qquad \alpha(f(t,B)) \leq \omega(t,\alpha(B)) \quad \text{for } t \in (0,a] \ , \ B \subset D \ ,$$

where $\omega: (0,a]\times\mathbb{R}_+ \to \mathbb{R}_+$ is such that to each $\varepsilon > 0$ there exist a $\delta > 0$, sequences (t_i) and (δ_i) with $t_i \to 0+$ and $\delta_i > 0$, and a sequence of functions ρ_i , continuous in $[t_i,a]$ with $\rho_i(t_i) \geq \delta t_i$, $D^-\rho_i(t) \geq \omega(t,\rho_i(t))+\delta_i$ and $\rho_i(t) \leq \varepsilon$ in $[t_i,a]$.
In other words, ω belongs to a rather general class of functions such that the initial value problem $\rho' = \omega(t,\rho)$ in $(0,a]$, $\rho(0) = 0$ has at most one solution with $\rho(t) = o(t)$ as $t \to 0+$; see § 3 and Deimling [51] for details. Particular cases have been considered earlier, for example by Ambrosetti [1] , Corduneanu [35] , Diaz/Bounds [52] , Goebel/Rzymowski [69] and Szufla [168] , where in most cases the results have been proved via the integral equation equivalent to (1) and the fixed point Lemma 2.3 .
Evidently, we could dispense with the uniform continuity of f if the following were true:
Given a sequence of continuously differentiable functions x_n with $\max_J|x_n'(t)| \leq c$ and $x_n(0) = x_0$ for every n , then

$$D^-\alpha(B_1(t)) \leq \alpha(B_1'(t)) \quad \text{in} \quad (0,a] \quad .$$

This seems to be an open question.

(v) Theorem 2.2 is taken from Deimling [51] . Cellina [27] has the special function $\omega(\rho) = L(\rho)\rho$, with

$$L(\rho) = \sup\{[\alpha(B)]^{-1}\alpha(f(J\times B)) : \alpha(B) \geq \rho\}$$

for $\rho > 0$, $L(0) = 0$ and $\int_{0+} \frac{d\rho}{L(\rho)\rho} = \infty$. One of the first existence theorems in this direction has been proved by Krasnoselskii/Krein [91]. They have $f = f_1+f_2$, where f_1 is Lipschitz with constant k and f_2 is compact, and they applied a weaker version of Lemma 2.3 , namely Krasnoselskii's fixed point theorem, where $(f_1+f_2)(C) \subset C$ is replaced by $f_1(C) + f_2(C) \subset C$. Therefore they had also to assume $b \cdot k < 1$.

(vi) A special case of (9) that has been considered e.g. by Browder [21,p.7] is the norm estimate $|f(t,x) - f(t,y)| \leq \omega(t,|x-y|)$. A familiar procedure to generalize such estimates is to take a Lyapunov-like function $V(t,x,y)$ instead of $|x-y|$. This idea also applies with respect to α :
Let $V: J \times \{B : B \subset D\} \to \mathbb{R}_+$ be such that V is continuous in t , $V(t,B) = 0$ iff $\bar{B}$ is compact, and

$$|V(t,B_1) - V(t,B_2)| \leq L|\alpha(B_1) - \alpha(B_2)| \quad .$$

Replace (9) by

$$(10) \qquad D^+V(t,B) \equiv \overline{\lim_{h\to 0+}} \; h^{-1}[V(t+h,B_h(f)) - V(t,B)] \leq \omega(t,V(t,B)) \quad ,$$

where $B_h(f) = \{x+hf(t,x) : x \in B\}$. It is also possible to take

$$D_-V(t,B) = \underline{\lim_{h\to 0-}} \; h^{-1}[V(t+h,B_h(f)) - V(t,B)]$$

in (10) . To prove a theorem corresponding to Theorem 2.1 , consider $\phi_k(t) = V(t,B_k(t))$ instead of $\alpha(B_k(t))$. A complete proof is given in Lakshmikantham [96] and Eisenfeld/Lakshmikantham [60] . Li [107] has the special case

$$\alpha(\{x-hf(t,x) : x \in B\}) \geq \alpha(B) - h\omega(t,\alpha(B)) \quad \text{for} \quad h > 0 \;,$$

and an example for f satisfying his condition but no estimate (2) .

(vii) Theorem 2.3 is from Deimling [51] . The same method has been used before by Deimling [47] and hereafter by Szufla [169] . Some particular cases have been considered before in Pulvirenti [144] and Vidossich [179]. A simple extension of Theorem 2.3 reads as follows.
Let the conditions of Theorem 2.1 or of Theorem 2.2 be satisfied ; $K \subset K_r(x_0)$ be compact and arcwise connected ;

$$r_1 = \max_K |x-x_0| \quad \text{and} \quad 0 < b < \min\{a,\frac{r-r_1}{c}\} \quad .$$

If S_y denotes the set of all solutions on $[0,b]$ of $x' = f(t,x)$, $x(0) = y$ then $\bigcup_{y\in K} S_y$ is a continuum of $C_X([0,b])$.
This result may be proved like Theorem 2.3 , considering now the ini-

tial value problems $x' = f_\lambda(t,x)$, $x(0) = y_\lambda \in K$, where y_λ is an arc in K connecting $x_1(0)$ and $x_2(0)$.
If $\dim X = \infty$, f is continuous and (1) has solutions then the set S of all solutions need not even be compact. An example of Cellina [27] is $X = l^\infty$, $x_0 = 0$, $f(x) = 2x/\sqrt{|x|}$ for $x \neq 0$ and $f(0) = 0$; the solutions are $x = 0$, $(t^2,0,\dots)$, $(0,t^2,0,\dots)$, ... See also Binding [12] .

(viii) Sometimes it may happen that the results given in this chapter work only after some change to an equivalent problem in particular if the differential equation (1) contains operators that are not continuous but have nice inverses. Let us illustrate this idea by an initial value problem that may be viewed as the abstract version of initial/boundary value problems for pseudo-parabolic equations like

$$\frac{\partial}{\partial t}\left(u - \frac{\partial^2 u}{\partial x^2}\right) - \frac{\partial^2 u}{\partial x^2} = u_t - u_{xx} - u_{xxt} = g(t,u(t,x))$$

in suitable Sobolev spaces. See e.g. Brill [20] and Showalter/Ting [162] and the references given there.
Let X,Y be real Banach spaces ; $A\colon D(A) \to Y$ and $B\colon D(B) \to Y$ be closed linear operators such that $D(B) \subset D(A) \subset X$, B is one to one and onto Y with $B^{-1}\colon Y \to D(B)$ compact. Let $J = [0,a]$ and $f\colon J\times X \to Y$ be continuous and consider the problem

$$(11) \qquad (Bu)' + Au = f(t,u) \quad , \quad u(0) = u_0 \in D(B) \; .$$

If u is a continuously differentiable function from J into D(B) satisfying (11) then $v = Bu$ is a solution of

$$(12) \qquad v' = -AB^{-1}v + f(t,B^{-1}v) \quad , \quad v(0) = Bu_0 \quad ,$$

and vice versa. We have $AB^{-1} \in L(Y)$ by the closed graph theorem, and $f(J\times B^{-1}(\Omega))$ relatively compact for every bounded $\Omega \subset Y$. Therefore (12) has a local solution, by Theorem 2.2 . This result is Brill's Theorem 3.2 .

§ 3 Conditions of dissipative type

Let X be a real Banach space, $J = [0,a] \subset \mathbb{R}$, $f: J\times X \to X$ and consider the initial value problem

$$(1) \qquad x' = f(t,x) \quad , \quad x(0) = x_0 \quad .$$

In case $X = \mathbb{R}$ it is well known that one-sided bounds like

$$(*) \qquad (f(t,x) - f(t,y))(x-y) \leq L(x-y)^2$$

give much better information about the behaviour of solutions than the stronger norm estimates like

$$|f(t,x) - f(t,y)| \leq L|x-y| \quad ,$$

where L has to be nonnegative. For instance, if $f(t,x) = -x$ and u,v are two solutions of the differential equation then (*) yields

$$|u(t) - v(t)| \leq |u(0) - v(0)|e^{-t}$$

while the norm condition gives us

$$|u(t) - v(t)| \leq |u(0) - v(0)|e^{t}$$

only. Evidently, (*) may be formulated as

$$(f(t,x) - f(t,y),x-y) \leq L|x-y|^2$$

in any inner product space, where $(\cdot,\cdot)$ denotes the inner product. In connexion with the Cauchy problem for systems of hyperbolic partial differential equations one was led to consider linear operators A with domain and range in a Hilbert space like L^2 satisfying $(Ax,x) \leq 0$, and such operators have been called dissipative since the energy of the corresponding system does not increase (see e.g. [141],[142]). On the other hand there are several initial value problems in partial differential equations where it is more natural to consider Banach spaces like L^p $(p \neq 2)$ rather than Hilbert spaces. This fact together with the natural aspiration to simulate inner-product techniques in connexion with other aspects of the theory of linear operators and the geometry of normed spaces has been the motivation to introduce a concept like inner product for arbitrary Banach spaces. This concept has been called semi-inner product, and one has studied operators which are dissipative with respect to such a semi-inner product ; see e.g. [111] , [112] . We shall define two semi-inner products $(\cdot,\cdot)_+$, $(\cdot,\cdot)_-$ which coincide with the given inner product in case X is a Hilbert space, and we consider conditions on f like

$$(2) \qquad (f(t,x) - f(t,y), x-y)_\pm \leq L|x-y|^2 ,$$

which will be called conditions of dissipative type. It will turn out that the initial value problem (1) has a unique solution if f is continuous and (2) is satisfied.

To avoid confusion, we want to mention that several authors prefer to call (2) a condition of accretive type, since many results have been formulated for operators A such that -A is dissipative, and operators of this type are called monotone in case of Hilbert spaces and accretive in case of general Banach spaces.

1. Duality maps and semi-inner products

Let X be a real Banach space and X^* its dual. Sometimes, we shall write $\langle x,x^*\rangle$ instead of $x^*(x)$, the value of x^* at x . As a result of the Hahn-Banach-Theorem, to each $x \in X$ there exists an $x^* \in X^*$ such that $|x^*| = 1$ and $x^*(x) = |x|$. Therefore,

$$\{x^* \in X : x^*(x) = |x|^2 , |x^*| = |x|\} \neq \emptyset$$

for every $x \in X$.

Definition 3.1. The mapping $F: X \to 2^{X^*}$, defined by

$$Fx = \{x^* \in X^* : x^*(x) = |x|^2 = |x^*|^2\} ,$$

is called the duality map of X . By means of F , the semi-inner products $(\cdot,\cdot)_\pm : X \times X \to \mathbb{R}$ are defined by

$$(x,y)_+ = \sup\{y^*(x) : y^* \in Fy\} \text{ and } (x,y)_- = \inf\{y^*(x) : y^* \in Fy\} .$$

It turns out that these semi-inner products have some properties of an inner product, but in general they are not linear with respect to the first or second argument, and useful continuity properties are available only for special classes of Banach spaces. We want to start with some facts about the duality map, and to this end let us recall some definitions.

The <u>w^*-topology</u> of X^* is generated by the neighborhoods of the origin

$$U_{\varepsilon,X_e} = \{x^*: |x^*(x)| < \varepsilon \text{ for } x \in X_e\} \quad ,$$

where $\varepsilon > 0$ and $X_e \subset X$ is finite.

The space X is called <u>strictly convex</u> if $x \neq y$ and $|x| = |y| = 1$ imply $|\lambda x+(1-\lambda)y| < 1$ for every $\lambda \in (0,1)$. The space X is called <u>uniformly convex</u> if to each $\varepsilon \in (0,2]$ there exists a $\delta(\varepsilon) > 0$ such that $|x| \leq 1$, $|y| \leq 1$ and $|x-y| \geq \varepsilon$ imply $|x+y| \leq 2(1-\delta(\varepsilon))$.

Given two topological spaces X , Y , a <u>multivalued map</u> $A: X \to 2^Y$ is called <u>continuous in x_0</u> if to every neighborhood V of Ax_0 there exists a neighborhood U of x_0 such that $A(U) \subset V$ (many authors call this property "upper semi-continuity") .

<u>Lemma 3.1</u>. Let $F: X \to 2^{X^*}$ be the duality map of X . Then

(i) Fx is convex and w^*-closed ; $F(\lambda x) = \lambda Fx$ for every $\lambda \in \mathbb{R}$.

(ii) If X^* is strictly convex then $F: X \to X^*$; in particular, $F = I$ in case X is Hilbert space.

(iii) F is continuous from X with the norm topology to X^* with the w^*-topology ("s-w^*-continuous") .

(iv) If X^* is uniformly convex then $F: X \to X^*$ is uniformly continuous on bounded subsets of X .

<u>Proof</u>. (i) $F(\lambda x) = \lambda Fx$ and the convexity of Fx follow immediately from Def. 3.1 . To prove that Fx is w^*-closed, let x^* be in the w^*-closure of Fx . Then, there is a net $(x_\mu^*) \subset Fx$ such that $x_\mu^*(y) \to x^*(y)$ for every $y \in X$; in particular $|x|^2 = x_\mu^*(x) \to x^*(x)$, and therefore $x^*(x) = |x|^2$; hence $|x^*| \geq |x|$. For every $y \in X$ with $|y| = 1$

$$x^*(y) \leq |\langle y,x^*-x_\mu^*\rangle| + |x_\mu^*| = |\langle y,x^* - x_\mu^*\rangle| + |x| \to |x| \quad .$$

Hence, we also have $|x^*| \leq |x|$, and therefore $x^* \in Fx$.

(ii) follows directly from Def. 3.1 and the Riesz-Representation-Theorem for functionals on a Hilbert space.

(iii) Suppose that F is not s-w^*-continuous in x_0 . Then, there are a neighborhood V of Fx_0 and sequences (x_n) , (x_n^*) such that $x_n \to x_0$, $x_n^* \in Fx_n$ and $x_n^* \notin V$. Let

$$M_k = \overline{\{x_n^* : n \geq k\}}^{w^*} \quad .$$

Since $|x_n^*| = |x_n|$ and $(|x_n|)$ is bounded, the M_k are uniformly bounded. By the Alaoglu-Theorem (see e.g. [43]) every bounded subset of X^* is w^*-relatively compact. Since also $M_k \supset M_{k+1}$, we have $\bigcap_{k\geq 1} M_k \neq \emptyset$ by the finite intersection property. Let $x_o^* \in \bigcap_{k\geq 1} M_k$. Since $M_k \cap V = \emptyset$ for every k, we have $x_o^* \notin V$. On the other hand, we shall show $x_o^* \in Fx_o$, a contradiction.

Consider the w^*-neighborhood $x_o^* + \{x^* : |x^*(x_o)| < 1/k\}$. Since $x_o^* \in M_k$ for all $k \geq 1$ we find an $x_{n_k}^*$ with $n_k \geq k$ in this neighborhood, and therefore a subsequence $(x_{n_k}^*)$ of (x_n^*) such that

$$|\langle x_o, x_{n_k}^* - x_o^*\rangle| < 1/k .$$

Hence,

$$|x_o^*(x_o) - |x_o|^2| \leq \frac{1}{k} + |\langle x_o - x_{n_k}, x_{n_k}^*\rangle| + |\langle x_{n_k}, x_{n_k}^*\rangle - |x_o|^2|$$

$$\leq \frac{1}{k} + |x_o - x_{n_k}||x_{n_k}| + ||x_{n_k}|^2 - |x_o|^2| \to 0 ,$$

and therefore $x_o^*(x_o) = |x_o|^2$. Since $x_o^* \in M_k$, we also have

$$|x_o^*| \leq \sup_{n\geq k}|x_n^*| = \sup_{n\geq k}|x_n| \to |x_o| \quad \text{as} \quad k \to \infty .$$

Therefore, $x_o^* \in Fx_o$.

(iv) Since "uniformly convex" implies "strict convex", we have $F:X\to X^*$. Suppose, F is not uniformly continuous on some bounded subset of X. Since F is homogenous we then find (x_n), (y_n) with $|x_n| = |y_n| = 1$, $|x_n - y_n| \to 0$ and $|Fx_n - Fy_n| \geq \varepsilon_o$ for some $\varepsilon_o > 0$ and all n. Hence, $|Fx_n + Fy_n| \leq 2(1-\delta(\varepsilon_o))$. On the other hand

$$|Fx_n + Fy_n| \geq \langle x_n, Fx_n + Fy_n\rangle = |x_n|^2 + |y_n|^2 + \langle x_n - y_n, Fy_n\rangle$$

$$\geq 2 - |x_n - y_n| \to 2 \quad \text{as} \quad n \to \infty ,$$

a contradiction.

q.e.d.

By means of Lemma 3.1 it is easy to derive the following properties of the semi-inner products that will be useful in the sequel.

<u>Lemma 3.2</u>. Let $(\cdot,\cdot)_+$ and $(\cdot,\cdot)_-$ be the semi-inner products from Def. 3.1 . Then

(i) $(x+y,z)_\pm \leq (x,z)_\pm + (y,z)_+$ and $|(x,y)_\pm| \leq |x||y|$;

$$(x+\alpha y,y)_\pm = (x,y)_\pm + \alpha|y|^2 \quad \text{for all } \alpha \in \mathbb{R} \quad ;$$

$$(\alpha x,\beta y)_\pm = \alpha\beta(x,y)_\pm \quad \text{for all } \alpha,\beta \in \mathbb{R} \text{ with } \alpha\beta \geq 0 \quad .$$

(ii) $(\cdot,\cdot)_+ = (\cdot,\cdot)_-$ if X^* is strictly convex, and both are equal to $(\cdot,\cdot)$ in case X is a Hilbert space.

(iii) $(x,y)_+ = \max\{y^*(x) : y^* \in Fy\}$ and $(x,y)_- = \min\{y^*(x) : y^* \in Fy\}$.

(iv) $(\cdot,\cdot)_+ : X\times X \to \mathbb{R}$ is upper semi-continuous.

(v) If X^* is uniformly convex then $(\cdot,\cdot)_\pm$ is uniformly continuous on bounded subsets of $X\times X$.

(vi) If $x: (a,b) \to X$ is differentiable at t and $\phi(t) = |x(t)|$ then $\phi(t)D^-\phi(t) \leq (x'(t),x(t))_-$.

Proof. (i) , (ii) and (v) follow immediately from Def. 3.1 and Lemma 3.1 .

(iii) By Lemma 3.1 (i) and Alaoglu's theorem, Fy is w^*-compact. Since $\phi: X^* \to \mathbb{R}$, defined by $\phi(y^*) = y^*(x)$ for x fixed, is w^*-continuous, ϕ attains its sup and inf on Fy .

(iv) Suppose, $(\cdot,\cdot)_+$ is not upper semi-continuous at some point (x_o,y_o). Then, there are an $\alpha > 0$; sequences (x_n) , (y_n) such that $x_n \to x_o$, $y_n \to y_o$; and, by (iii) , a sequence (y_n^*) with $y_n^* \in Fy_n$, such that $<x_n,y_n^*> \geq <x_o,y^*>+\alpha$ for every n and $y^* \in Fy_o$. Hence, $<x_o,y_n^*-y^*> \geq \alpha/2$ for every $n \geq n_o$ and $y^* \in Fy_o$, contradicting the s-w^*-continuity of F .

(vi) Let $x^* \in Fx(t)$. Then $<x(t)-x(t-h),x^*> \geq \phi^2(t)-\phi(t)\phi(t-h)$. We divide by $h > 0$ and let $h \to 0+$ to obtain the assertion.

q.e.d.

For some spaces X it is easy to determine Fx explicitely. Let us consider

Example 3.1. (i) Let

$$l^p = \{x \in \mathbb{R}^{\mathbb{N}} : |x| = (\sum_{i\geq 1} |x_i|^p)^{1/p} < \infty\}$$

for $1 < p < \infty$. Then $(l^p)^* = l^q$, where $p^{-1} + q^{-1} = 1$. Since these spaces l^p are strictly convex, $F: l^p \to l^q$ is given by $F(0) = 0$,

$$(Fx)_i = |x|^{2-p}|x_i|^{p-1}\operatorname{sgn} x_i \quad \text{for } i \geq 1 \text{ and } x \neq 0 \ .$$

Therefore,

$$(x,y)_+ = (x,y)_- = |y|^{2-p} \sum_{i\geq 1} x_i |y_i|^{p-1} \operatorname{sgn} y_i \quad .$$

(ii) For $L^p(\Omega,\mathfrak{A},\mu)$ with $1 < p < \infty$ we have the corresponding representation $F(0) = 0$,

$$(Fx)(\omega) = |x|^{2-p}|x(\omega)|^{p-1}\operatorname{sgn} x(\omega) \quad \text{for } x \neq 0 \quad ,$$

and

$$(x,y)_+ = (x,y)_- = |y|^{2-p} \int_\Omega x(\omega)|y(\omega)|^{p-1}\operatorname{sgn} y(\omega)d\mu(\omega) \quad .$$

(iii) Consider l^1 . Since $(l^1)^* = l^\infty$, we have

$$Fx = \{z \in l^\infty : \sum_{i\geq 1} x_i z_i = |x|^2 \text{ and } \sup_i |z_i| = |x| = \sum_{i\geq 1}|x_i|\} \quad .$$

Suppose $|x_j| \neq 0$, and let $\alpha_i = z_i|x|^{-1}$. Then, $|\alpha_i| \leq 1$ and

$$\sum_{i\geq 1} |x_i|\alpha_i \operatorname{sgn} x_i = \sum_{i\geq 1} |x_i| \quad .$$

Therefore, $\alpha_j \operatorname{sgn} x_j = 1$. Hence,

$$Fx = \{z \in l^\infty : z_i = |x| \operatorname{sgn} x_i \text{ for } i \text{ with } x_i \neq 0 \text{ , } z_i \in \mathbb{R} \text{ arbitrary with } |z_i| \leq |x| \text{ if } x_i = 0\} \quad .$$

Let $A = \operatorname{supp} y = \{i : y_i \neq 0\}$. Then, it is easy to see that

$$(x,y)_+ = |x|\,|y| - |y| \sum_{i\in A}(|x_i| - x_i \operatorname{sgn} y_i)$$

and

$$(x,y)_- = -|x|\,|y| + |y| \sum_{i\in A}(|x_i| + x_i \operatorname{sgn} y_i) \quad .$$

In particular, $(x,y)_- = (x,y)_+$ if and only if either $y = 0$ or $\operatorname{supp} x \subset \operatorname{supp} y$.

2. Uniqueness

Consider the initial value problem (1) , and let us assume that

$$(3) \qquad (f(t,x)-f(t,y),x-y)_- \leq \omega(t,|x-y|)|x-y| \quad ,$$

where ω is essentially such that $\rho(t) \equiv 0$ is the only solution of $\rho' = \omega(t,\rho)$ with $\rho(t) = o(t)$ as $t \to 0$. To be more precise, we assume that ω belongs to the following general class U of uniqueness functions.

Definition 3.2. A function ω: $(0,a]\times\mathbb{R}_+ \to \mathbb{R}$ is said to be of class U if to each $\varepsilon > 0$ there exist $\delta > 0$, a sequence $t_i \to 0+$ and a sequence of continuous functions ρ_i: $[t_i,a] \to \mathbb{R}_+$ with

$$\rho_i(t_i) \geq \delta t_i \ , \ D^-\rho_i(t) > \omega(t,\rho_i(t)) \ , \ 0 < \rho_i(t) \leq \varepsilon \text{ in } (t_i,a] \quad .$$

It is easy to see that all well known uniqueness functions ω are of class U , for example, the Lipschitz condition $\omega(t,\rho) = L\rho$ and the Nagumo condition $\omega(t,\rho) = \rho/t$. In the latter case, consider $\delta = \varepsilon e^{-a}/a$, $\rho_\varepsilon(t) = t(\varepsilon/a)\exp(t-a)$, any sequence $t_i \to 0+$ and $\rho_i = \rho_\varepsilon\big|[t_i,a]$; see Remark (ii) .

Theorem 3.1. Let X be a real Banach space, $J = [0,a] \subset \mathbb{R}$, $D \subset X$ and $x_0 \in D$. Let f: $J\times D \to X$ satisfy (3) for $t \in (0,a]$ and $x,y \in D$, with $\omega \in U$. Then, (1) has at most one solution in J .

Proof. Let x,y be solutions in J and $\phi(t) = |x(t)-y(t)|$. By (2) and Lemma 3.2 (vi) , we have $\phi(t)D^-\phi(t) \leq \omega(t,\phi(t))\phi(t)$ in $(0,a]$, and $\phi(t) = o(t)$ as $t \to 0+$, since

$$\frac{\phi(t)}{t} = \left|\frac{x(t)-x_0}{t} - \frac{y(t)-x_0}{t}\right| \to |f(0,x_0)-f(0,x_0)| = 0 \quad .$$

Given $\varepsilon > 0$, we choose $\delta > 0$ from Def. 3.2 ; $t_0 > 0$ such that

$$\phi(t) \leq \frac{\delta}{2}t \qquad \text{for} \qquad t \leq t_0 \qquad ;$$

$t_i < t_0$ and ρ_i from Def. 3.2 . Hence, $\phi(t_i) < \rho_i(t_i)$. Suppose, there is a first time $t^* > t_i$ such that $\phi(t^*) = \rho_i(t^*)$. Then, $\phi(t^*) > 0$ and therefore

$$D^-\phi(t^*) \leq \omega(t^*,\phi(t^*)) = \omega(t^*,\rho_i(t^*)) < D^-\rho_i(t^*) \quad ,$$

which is impossible since $\phi(t) < \rho_i(t)$ in $[t_i,t^*)$. Therefore, $\phi(t) \leq \varepsilon$ in J for every $\varepsilon > 0$.

q.e.d.

Evidently, Theorem 3.1 remains true if $x' = f(t,x)$ holds in $(0,a]$ only, but x is required to satisfy

$$\lim_{t\to 0+} (x(t)-x_0)/t = x_1$$

for some given x_1 .

3. Local existence

Let U_1 be the class of all $\omega \in U$ such that the functions $\rho_i(t)$ from Def. 3.2 satisfy in addition

$$D^-\rho_i(t) \geq \omega(t,\rho_i(t))+\delta_i \quad \text{for some } \delta_i > 0$$

(for instance, ω Lipschitz, Nagumo,...) .

Theorem 3.2. Let X be a real Banach space, $J = [0,a] \subset \mathbb{R}$, $D = \bar{K}_r(x_o) \subset X$, $f\colon J\times D \to X$ continuous, $|f(t,x)| \leq c$ on $J\times D$. Let f satisfy condition

$$(3) \qquad (f(t,x)-f(t,y),x-y)_- \leq \omega(t,|x-y|)|x-y| \quad \text{for} \quad t \in (0,a] \;;\; x,y \in D \;,$$

with $\omega \in U_1$. Let $b < \min\{a,r/c\}$. Then (1) has exactly one solution on $[0,b]$.

Proof. By Theorem 1.1 , we have approximate solutions x_n on $[0,b]$ such that

$$x_n'(t) = f(t,x_n(t))+y_n(t) \quad , \quad x_n(0) = x_o \;, \quad |y_n(t)| \leq 1/n \quad .$$

Let $z(t) = x_n(t)-x_m(t)$ and $\phi(t) = |z(t)|$. By Lemma 3.2 , we have

$$(*) \qquad \begin{aligned} \phi(t)D^-\phi(t) &\leq (z'(t),z(t))_- \\ &\leq \omega(t,\phi(t))\phi(t)+|y_n(t)-y_m(t)|\phi(t) \\ &\leq \omega(t,\phi(t))\phi(t) + (\tfrac{1}{n}+\tfrac{1}{m})\phi(t) \quad \text{in } (0,b] \;. \end{aligned}$$

Since f is continuous in $(0,x_o)$ and $|x_n(t)-x_o| \leq (c+1)t$, it is easy to see that to each $\eta > 0$ there exists $t_\eta > 0$ such that

$$\phi(t) \leq (\tfrac{1}{n}+\tfrac{1}{m}+\eta)t \quad \text{for } t \in [0,t_\eta] \quad .$$

Since $\omega \in U_1$, given $\varepsilon > 0$ we choose $\delta > 0$ from Def. 3.2 , $\eta > 0$ and n_o such that $1/n + 1/m + \eta < \delta$ for n and $m \geq n_o$, $t_\eta > 0$, $t_i < t_\eta$ and the function ρ_i . Now, we choose $n_1 \geq n_o$ such that $1/n + 1/m < \delta_i$ for $n,m \geq n_1$. As in the proof of Theorem 3.1 , (*) implies

$$\phi(t) \leq \rho_i(t) \leq \varepsilon \quad \text{for } n,m \geq n_1 \quad .$$

Hence, (x_n) is a Cauchy sequence in $C_X([0,b])$. By a standard argument, $\lim_{n\to\infty} x_n(t)$ is a solution of (1) , and uniqueness follows from Theorem 3.1.

q.e.d.

4. Global existence

Up to now, we have only been concerned with local existence (Theorems 2.1 , 2.2 and 3.2) . To ensure that all solutions of the initial value problem (1) exist on the whole interval under consideration, we have to impose some growth conditions on f .

Theorem 3.3. Let X be a real Banach space, $J = [0,a]$ or $[0,a)$ with $a \leq \infty$, $f: J\times X \to X$ continuous, and consider the following hypotheses

(i) The problem $x' = f(t,x)$, $x(t_1) = x_1$ has a local solution in $t \geq t_1$, for every $(t_1,x_1) \in [0,a)\times X$.

(ii) $(f(t,x),x)_- \leq \tilde{\omega}(t,|x|)|x|$ in $J\times X$, where $\tilde{\omega}: J\times \mathbb{R}_+ \to \mathbb{R}$ is continuous and such that the maximal solution ρ^* of $\rho' = \tilde{\omega}(t,\rho)$, $\rho(0) = |x_o|$ exists on J and is nonnegative.

(iii) f maps bounded subsets of $J\times X$ into bounded sets.

If (i)-(iii) are satisfied, then (1) has a solution on J .

Proof. Let x be a solution of (1) on some interval $[0,\alpha_x)$, and $\phi(t) = |x(t)|$. Then $\phi(t)D^-\phi(t) \leq \tilde{\omega}(t,\phi(t))\phi(t)$ in $(0,\alpha_x)$ and $\phi(0) = \rho^*(0)$. Since ρ^* may be approximated from above, uniformly on compact intervals, we obtain $\phi(t) \leq \rho^*(t)$.
Let S be the set of solutions x of (1) , existing on $J_x = [0,\alpha_x)$ and satisfying $|x(t)| \leq \rho^*(t)$. S is not empty. By "$x \leq y \iff J_x \subset J_y$ and $y(t) = x(t)$ on J_x" , we have defined a partial ordering $\leq$ on S . If S_o is a chain in S , then y , defined by

$$J_y = [0,\sup_{x\in S_o} \alpha_x) \quad \text{and} \quad y|_{J_x}(t) = x(t)$$

is an upper bound for S_o . Hence, by Zorn's theorem, S has a maximal element u . Suppose, $\alpha_u < a$. Then, $\{u(t) : t < \alpha_u\}$ is bounded, and by (iii) , $\{u'(t) : t < \alpha_u\}$ is bounded too. Therefore, $u_o = \lim_{t\to\alpha_u} u(t)$ exists. We define $u(\alpha_u)$ to be u_o . By (i) , u has an extension $\bar{u}$ to $[0,\alpha)$ for some $\alpha > \alpha_u$, and $\bar{u}$ satisfies $|\bar{u}(t)| \leq \rho^*(t)$ in $[0,\alpha)$. Hence, u is not maximal ; a contradiction. Therefore, $\alpha_u = a$, and we are done in case $J = [0,a)$. If $J = [0,a]$, we repeat the argument given for $t \to \alpha_u$, to see that $\lim_{t\to a} u(t)$ exists.

q.e.d.

Obviously, (iii) is satisfied if the strong version $|f(t,x)| \leq \tilde{\omega}(t,|x|)$ of (ii) holds. But in that case ρ^* is increasing, while ρ^* in (ii) may decrease. For autonomous equations with dissipative right hand side, we have

Theorem 3.4. Let X be a real Banach space, $J = [0,a]$ or $[0,a)$ with $a \leq \infty$, $f: X \to X$ continuous and

$$(f(x)-f(y),x-y)_- \leq \omega(|x-y|)|x-y| \quad ,$$

where $\omega: \mathbb{R}_+ \to \mathbb{R}$ is continuous and such that $\rho^*(t) \equiv 0$ is the maximal solution of the initial value problem $\rho' = \omega(\rho)$ in J, $\rho(0) = 0$. Then, (1) has a unique solution on J, for every $x_0 \in X$.

Proof. Obviously, ω is of class U_1. Therefore, (1) has a unique solution x in a maximal interval $[0,\alpha)$ with $\alpha \leq a$. As in the proof of Theorem 3.3, we have to show that $\alpha < a$ implies the existence of $\lim_{t\to\alpha} x(t)$. Let $h > 0$ and $\phi(t) = |x(t+h)-x(t)|$ for $t < \alpha-h$. We have $\phi(0) = |x(h)-x_0|$ and $\phi(t)D^-\phi(t) \leq \omega(\phi(t))\phi(t)$ for $t > 0$. Since $\omega \in U_1$, to $\varepsilon > 0$ there exists $\delta > 0$ such that $\phi(0) < \delta$ implies $\phi(t) \leq \varepsilon$. Therefore, since x is continuous, we find $\delta_1(\varepsilon) > 0$ such that $h \leq \delta_1(\varepsilon)$ implies $\phi(0) < \delta$ and therefore $\phi(t) \leq \varepsilon$ in $[0,\alpha-h)$. Hence, $\lim_{t\to\alpha} x(t)$ exists.

q.e.d.

5. Excursion to stochastic differential equations

In many problems of natural sciences a given system is subject to random influences that must be taken into account in the appropriate mathematical models. For example, one may think of falsified measurement indications of mirror galvanometers, gyroscopic devices etc. produced by random mechanical shocks. Here, the differential equations describing the forced oscillations of the system contain forcing terms which are stochastic rather than deterministic variables. Therefore the solutions of the equations will be stochastic processes rather than deterministic functions.
Thus, we are led to consider a probability measure space $(\Omega,\mathfrak{A},\mu)$, an interval $J = [0,a] \subset \mathbb{R}$, functions $f: J\times\mathbb{R}^n\times\Omega \to \mathbb{R}^n$ and $x_0: \Omega \to \mathbb{R}^n$, and the initial value problem

$$(4) \qquad x'(t,\omega) = f(t,x(t,\omega),\omega) \qquad \text{for } t \in J \,,\ \omega \in \Omega$$

$$(5) \qquad x(0,\omega) = x_0(\omega) \qquad \text{for } \omega \in \Omega \,.$$

A natural way to define a (local) "solution" to (4), (5) is to look for a function $x: [0,\delta]\times\Omega \to \mathbb{R}^n$, for some $\delta \in (0,a]$, such that $x(\cdot,\omega)$ is differentiable in $[0,\delta]$ and (4), (5) are satisfied for almost all $\omega \in \Omega$. Since we want to obtain information on $x(t,\cdot)$ in terms like expectation, variance etc., x should also be a stochastic process, i.e. $x(t,\cdot)$ should be $\mathfrak{A}$-measurable for each $t \in [0,\delta]$. Let us call x a <u>solution of the first kind</u> if it has all these properties.
Now, suppose that the function x_0 belongs to $L^p(\Omega,\mathfrak{A},\mu)$ for some $p \geq 1$. Then it is also natural to look for p-th order stochastic processes x, i.e. for functions $x: [0,\delta]\times\Omega \to \mathbb{R}^n$ such that $x(t,\cdot) \in L^p(\Omega,\mathfrak{A},\mu)$ for all $t \in [0,\delta]$, and to consider (4), (5) as an initial value problem

(6) $$u' = F(t,u) \quad , \quad u(0) = x_0$$

in the Banach space $X = L^p(\Omega,\mathfrak{A},\mu)$, where $F(t,u)(\omega) = f(t,u(\omega),\omega)$ for $\omega \in \Omega$. In case (6) has a solution u, we call x, defined by

$$x(t,\omega) = u(t)(\omega) \quad ,$$

a <u>solution of the second kind</u> to (4), (5).
Suppose for example that f does not depend on x but is only known to be a continuous second order process, i.e. $p = 2$ and $t \to f(t,\cdot)$ is continuous, and let $x_0 \in L^2(\Omega,\mathfrak{A},\mu)$. Then (4), (5) has a unique solution of the second kind, given by

$$x(t,\omega) = x_0(\omega) + \int_0^t f(s,\omega)ds \quad ,$$

but we can not say anything about existence of solutions of the first kind. If, however, we know more about f, for instance that $f(\cdot,\omega)$ is continuous for almost all $\omega \in \Omega$, then the solution of the second kind is also of first kind.
The following existence theorem, where we let $n = 1$ and $p = 2$ for simplicity, is a simple consequence of Theorem 3.2 and Theorem 3.3.

<u>Theorem 3.5</u>. Let $(\Omega,\mathfrak{A},\mu)$ be a probability measure space ; $X = L^2(\Omega,\mathfrak{A},\mu)$ and

$$\|x\|^2 = \int_\Omega |x(\omega)|^2 d\mu(\omega) \qquad \text{for } x \in X \quad ;$$

$J = [0,a]$ or $[0,a)$ with $a \leq \infty$; $x_0 \in X$. Let the functions $g,h: J\times\mathbb{R}\times\Omega \to \mathbb{R}$ be such that

(i) $g(t,x,\cdot)$, $h(t,x,\cdot)$ are measurable for all $(t,x) \in J\times\mathbb{R}$; $g(\cdot,\cdot,\omega)$ and $h(\cdot,\cdot,\omega)$ are continuous for almost all $\omega \in \Omega$.

(ii) $$|g(t,x,\omega) - g(t,y,\omega)| \leq L(t)|x-y|$$

for some continuous function $L: J \to \mathbb{R}_+$, all $t \in J$ and $x,y \in \mathbb{R}$, and almost all $\omega \in \Omega$; $|g(t,0,\omega)| \leq c(t)\phi(\omega)$ with $c: J \to \mathbb{R}_+$ continuous and $\phi \in X$.

(iii) The function $h(t,\cdot,\omega)$ is monotone decreasing, for all $(t,\omega) \in J\times\Omega$, and there exist a continuous function $d: J \to \mathbb{R}_+$ and $\psi \in X$ such that

$$(7) \qquad |h(t,x,\omega)| \leq d(t)(\psi(\omega)+|x|) \qquad \text{in } J\times\mathbb{R}^1\times\Omega \quad .$$

Then problem (4) , (5) with $f = g+h$ has a unique global solution x of the second kind (second order process) .

Proof. Let $G(t,u)(\omega) = g(t,u(\omega),\omega)$, $H(t,u)(\omega) = h(t,u(\omega),\omega)$ and $F = G+H$. Since $|g(t,x,\omega)| \leq c(t)\phi(\omega) + L(t)|x|$ and g is Lipschitz in x , we have $G: J\times X \to X$ and $\|G(t,u)-G(t,v)\| \leq L(t)\|u-v\|$. The estimate (7) implies $H: J\times X \to X$, and since $h(t,\cdot,\omega)$ is nonincreasing, we have $(H(t,u)-H(t,v),u-v) \leq 0$. Therefore, $F: J\times X \to X$ and

$$(F(t,u)-F(t,v),u-v) \leq L(t)\|u-v\|^2 \quad \text{for } t \in J \text{ and } u,v \in X \; .$$

Let us prove that H is continuous. Let $t_n \to t_o$ and $\|u_n-u_o\| \to 0$. Consider any subsequence (u_m) such that $u_m \to u_o$ [μ-a.e.] . Then

$$h(t_m,u_m(\omega),\omega) \to h(t_o,u_o(\omega),\omega) \; [\mu\text{-a.e.}] \quad ,$$

by condition (i) . Furthermore

$$(8) \qquad \int_A |h(t_m,u_m(\omega),\omega)|^2 d\mu \leq d^2(t_m)\{\int_A |\psi(\omega)|^2 d\mu + \int_A |u_m(\omega)|^2 d\mu\}$$

for $A \in \mathcal{A}$. Since $\|u_m-u_o\| \to 0$, we have

$$\int_A |u_m(\omega)|^2 d\mu \to 0 \qquad \text{as} \qquad \mu(A) \to 0 \quad ,$$

uniformly with respect to m .
Since d is continuous, $(d(t_m))$ is bounded, and therefore (8) implies

$$\int_A |h(t_m,u_m(\omega),\omega)|^2 d\mu \to 0 \qquad \text{as} \qquad \mu(A) \to 0 \quad ,$$

uniformly with respect to m . Thus, we may apply Vitali's convergence theorem (see e.g. [74,p.203]) to conclude that

$$\|H(t_m,u_m)-H(t_o,u_o)\| \to 0 \qquad \text{as} \qquad m \to \infty \quad .$$

Hence, H must be continuous. The continuity of G follows along the same lines, and therefore F is continuous. Finally,

$$\|F(t,u)\| \leq c(t)\|\phi\| + d(t)\|\psi\| + \{L(t) + d(t)\}\|u\| \quad .$$

Therefore, the initial value problem (6) has a unique solution on J .

q.e.d.

6. Excursion to continuous accretive operators

Let us indicate how the existence theorems of this chapter may be applied in the study of accretive operators i.e. of mappings T from $D(T) \subset X$ into X such that

$$(Tx-Ty,x-y)_+ \geq 0 \quad \text{for all } x,y \in D(T) .$$

A basic result is the following

Theorem 3.6. Let X be a real Banach space, $T: X \to X$ continuous and $(Tx-Ty,x-y)_+ \geq c|x-y|^2$ for some $c > 0$. Then T is a homeomorphism from X onto X .

Proof. Obviously, T is one to one, and $T^{-1}: R(T) \to X$ is Lipschitz, since

$$c|T^{-1}x - T^{-1}y|^2 \leq (x-y,T^{-1}x - T^{-1}y)_+ \leq |T^{-1}x - T^{-1}y||x-y| \quad .$$

Hence, we have to show $R(T) = X$. If $y \in X$ is fixed, then $T(\cdot)-y$ has the same properties as T . Therefore, it is sufficient to prove $0 \in R(T)$. Consider the initial value problem

$$(9) \qquad u' = -Tu \quad , \quad u(0) = x \in X \quad .$$

Since $f = -T$ satisfies $(f(x)-f(y),x-y)_- \leq -c|x-y|^2$, (9) has a unique solution $u(t,x)$ on $[0,\infty)$, and $|u(t,x)-u(t,y)| \leq e^{-ct}|x-y|$. Let $p > 0$ and define $U: X \to X$ by $Ux = u(p,x)$. Since $|Ux-Uy| \leq e^{-cp}|x-y|$, U has a unique fixed point x_0 . Since $u(p,x_0) = x_0 = u(0,x_0)$, the solution $u(\cdot,x_0)$ is p-periodic, i.e. $u(t+p,x_0) = u(t,x_0)$ for each $t \geq 0$. Hence,

$$|u(t,x_0) - x_0| = |u(t+p,x_0) - x_0| = |Uu(t,x_0) - Ux_0|$$
$$\leq e^{-cp}|u(t,x_0) - x_0| \quad .$$

This implies $u(t,x_0) \equiv x_0$, $u'(t,x_0) \equiv 0$, and therefore $Tx_0 = 0$.

q.e.d.

Let us mention two simple consequences for dissipative operators, i.e. operators such that $-A$ is accretive. If $A \in L(X)$ is dissipative then every $\lambda > 0$ is in the resolvent set of A, since $T = \lambda I - A$ satisfies Theorem 3.5 with $c = \lambda$.
Secondly, a continuous dissipative operator $A: X \to X$ is maximal dissipative, i.e. there is no dissipative not necessarily linear $\tilde{A}: X \to X$ such that graph A is a proper subset of graph $\tilde{A}$; in other words:

$$(Ax-y_o, x-x_o)_- \leq 0 \quad \text{for all } x \in X \quad \text{implies} \quad y_o = Ax_o \ .$$

In fact, since $I-A$ is onto, we have $x_1 - Ax_1 = x_o - y_o$ for some $x_1 \in X$, hence $|x_1 - x_o|^2 \leq 0$ and therefore $y_o = Ax_o$.
If $T: X \to X$ is continuous and accretive then $T + \varepsilon I$ is a homeomorphism from X onto X, for every $\varepsilon > 0$. In order to show how this observation may be applied, let us prove

Theorem 3.7. Let X be a real Banach space, $T: X \to X$ continuous and accretive. Suppose in addition that T satisfies one of the following hypotheses

(i) $|Tx| \to \infty$ in case $|x| \to \infty$.

(ii) To each $y \in X$ there exists an $r = r(y) > 0$ such that $(Tx-y,x)_+ \geq 0$ for $|x| \geq r$.

Then $T(X)$ is dense in X.

Proof. Let $y \in X$. By Theorem 3.6 there exists x_n such that $Tx_n + \frac{1}{n}x_n = y$. Since T is accretive, we have

$$0 \leq (Tx_n - T(0), x_n)_+ = (y - T(0) - \tfrac{1}{n}x_n, x_n)_+$$
$$\leq -\tfrac{1}{n}|x_n|^2 + |y - T(0)|\,|x_n| \qquad ,$$

hence $|Tx_n| \leq |y| + |y - T(0)|$ for all n. If we assume (i), then (x_n) must be bounded too. If, however, (ii) holds then we have $|x_n| < r$, since $|x_n| \geq r$ would imply $0 \geq \frac{1}{n}|x_n|^2$. This shows that (x_n) is bounded in both cases, and therefore $Tx_n \to y$, i.e. $y \in \overline{T(X)}$.

q.e.d.

For example, condition (ii) is satisfied if T is "coercive", i.e. if $(Tx,x)_+/|x| \to \infty$ as $|x| \to \infty$. In the proof of Theorem 3.6 we have been able to apply Banach's fixed point theorem since the operators $U(t)$ had been strict contractions. If T is only accretive (i.e. $c = 0$) then $U(t)$ is only nonexpansive, i.e. $|U(t)x - U(t)y| \leq |x-y|$. For nonexpansive mappings one has, among others, the following fixed point theorem ; see [21, Theorem 8.7] .

Lemma 3.3. Let X be a real uniformly convex Banach space ; $C \subset X$ closed bounded convex, $(T_\lambda)_{\lambda \in \Lambda}$ a commuting family of nonexpansive maps $T_\lambda : C \to C$. Then the T_λ have a common fixed point.

In the proof of the following simple result we want to illustrate how this Lemma can be applied.

Theorem 3.8. Let X be a real uniformly convex Banach space, $T: X \to X$ a continuous accretive mapping such that condition (ii) in Theorem 3.7 holds. Then $T(X) = X$.

Proof. Consider a fixed $y \in X$, $\tilde{T}x = Tx - y$ for $x \in X$ and $r = r(y)$ from (ii) . The initial value problem $u' = -\tilde{T}u$, $u(0) = x \in \bar{K}_r(0)$ has a unique solution $U(t)x$ on $[0,\infty)$. For $\phi(t) = |U(t)x|$ we obtain

$$\phi(t)D^-\phi(t) \leq 0 \quad \text{if} \quad \phi(t) \geq r \quad ,$$

by (ii) . Hence, $U(t): \bar{K}_r(0) \to \bar{K}_r(0)$. Furthermore, $U(t)$ is nonexpansive and $U(t)U(s) = U(s)U(t)$ for $s,t \geq 0$. By Lemma 3.3 , there exists an $x_0 \in \bar{K}_r(0)$ such that $U(t)x_0 \equiv x_0$. Therefore $\tilde{T}x_0 = 0$, i.e. $Tx_0 = y$.

q.e.d.

7. Remarks

(i) The theorems of this chapter are also true for complex Banach spaces, where we define

$$(x,y)_+ = \max\{\mathrm{Re}\; y^*(x) : y^* \in Fy\}$$

and

$$(x,y)_- = \min\{\mathrm{Re}\; y^*(x) : y^* \in Fy\} \quad .$$

(ii) More details concerning duality mappings may be found in the books of Barbu [7] , Browder [21] , Cioranescu [33] and Pascali [136] and the references given there. Theorems 3.1 to 3.4 are taken from Deimling [50] . Related uniqueness theorems are in Goldstein [70] , and more examples for functions ω of class U may be found in the books of Lakshmikantham/Leela [97] and Walter [187,p.81] . Earlier versions of Theorems 3.2 - 3.4 are in Vidossich [175] ; see also the survey article of Nemytskii/Vainberg/Gusarova [130,pp.173/74] .

(iii) As in Remark (vi) of § 2 , it is easily seen that theorems of the above type may also be proved by means of conditions on f involving Lyapunov-like functions $V(t,x,y)$.
Let $D = \overline{K}_r(x_0) \subset X$ and $V: J\times D\times D \to \mathbb{R}_+$ be such that $V(t,x,x) = 0$, $V(t,x,y) > 0$ if $x \neq y$, $|V(t,x,y) - V(t,\overline{x},\overline{y})| \leq L(|x-\overline{x}|+|y-\overline{y}|)$, and $V(t,x_n,y_n) \to 0$ implies $|x_n-y_n| \to 0$ as $n \to \infty$. Instead of condition (3), consider

$$(10) \quad \overline{\lim_{h\to 0+}}\, h^{-1}(V(t+h,x+hf(t,x),y+hf(t,y)) - V(t,x,y)) \leq \omega(t,V(t,x,y))$$

for $t \in J$ and $x,y \in D$. To prove the existence theorem, consider $\phi(t) = V(t,x_n(t),x_m(t))$ instead of $\phi(t) = |x_n(t)-x_m(t)|$, where (x_n) is a sequence of approximate solutions.
Results of this kind may be found e.g. in Lakshmikantham [96] , Martin [118] , Murakami [127] .
For example, let us choose $V(t,x,y) = |x-y|$. Then, (10) becomes

$$(11) \quad \overline{\lim_{h\to 0+}}\, h^{-1}(|x-y+h(f(t,x) - f(t,y))| - |x-y|) \leq \omega(t,|x-y|) \quad .$$

Multiplying (11) by $|x-y|$, we obtain

$$x^*(f(t,x) - f(t,y)) \leq \omega(t,|x-y|)|x-y| \quad \text{for every } x^* \in F(x-y) \quad ,$$

i.e. condition (3) with $(\ ,\ )_+$ instead of $(\ ,\ )_-$.

(iv) In § 2 we have considered compact perturbations of Lipschitz right hand sides, by means of measures of noncompactness. It is also possible to prove existence theorems for compact perturbations of dissipative maps f . To this end, Cellina [28] introduced the concept of α-dissipative mappings. A map $f: J\times D \to X$ is called α-dissipative if for every $\varepsilon > 0$ there exists a finite covering (Ω_i) of $J\times D$ such that $(f(t,x) - f(\overline{t},\overline{x}),x-\overline{x})_- \leq \varepsilon$ whenever (t,x) and $(\overline{t},\overline{x})$ belong to the same Ω_i .
Obviously, a function f satisfying (3) with $\omega \equiv 0$ is α-dissipative. If f is compact, there is a finite open covering $(\tilde{\Omega}_i)$ of $f(J\times D)$ such that $\mathrm{diam}(\tilde{\Omega}_i) \leq \varepsilon/\mathrm{diam}(J\times D)$; hence, we see that f is α-dissipative by setting $\Omega_i = f^{-1}(\tilde{\Omega}_i)$. Moreover, f_1+f_2 is α-dissipative if f_1 and f_2 are.
Cellina has shown that problem (1) has a local solution provided f is continuous and α-dissipative, and X^* is uniformly convex (in order to have F uniformly continuous on bounded sets; see (iv) in Lemma 3.1) .

The prolongability of local solutions for autonomous equations with α-dissipative f has been studied in Cellina/Pianigiani [30] . Li [107] has generalized Cellina's existence theorem assuming only $(f(t,x) - f(\bar{t},\bar{x}),x-\bar{x})_- \leq L|x-\bar{x}|^2+\varepsilon|x-\bar{x}|$ if $(t,x),(\bar{t},\bar{x})$ are in the same Ω_i . Furthermore, he shows that this condition implies

$$\alpha(\{x-hf(t,x) : x \in B\}) \geq (1-Lh)\alpha(B) \quad \text{for} \quad h \geq 0 \quad ;$$

cp. Remark (vi) in § 2 . Hence, if f is also uniformly continuous, then X^* need not be uniformly convex.
It seems to be open wether (1) has a local solution if $f = f_1+f_2$, f_1 compact, f_2 continuous and dissipative, and X arbitrary. Martin [116] has another additional condition: either f_1 or f_2 is uniformly continuous.

(v) The "stochastic" Theorem 3.5 is known for Lipschitzian right hand sides (i.e. $h(t,x,\omega) \equiv 0$) ; see Bharucha-Reid [11,Chap.6] and the references given there. For more results like those in sect. 6 we refer to chap. 8 and chap. 10 in the book of Browder [21] and to the Remarks that will be given at the end of § 4 .

§ 4 Solutions in closed sets

In the preceding chapters we have always assumed that the initial point x_0 is an inner point of the domain of f . However there are several basic problems where one is interested in solutions through boundary points, in particular if the domain of f has no inner points at all.

For example, global existence, stability, existence of periodic solutions and theorems on differential inequalities are intimately connected with the flow invariance of a certain subset D in the domain of f , i.e. with the question whether every solution starting in D remains in D as long as it exists. If we start at an inner point of D we may arrive at the boundary after a short time. If the interior of D is empty we even have to worry about local existence. Moreover, we have seen that in infinite dimensional spaces we need more than continuity of f to guarantee local existence. Now, for example, it may happen that f is compact in a set D without interior points, but not on a full neighborhood of x_0 .

We shall consider a real Banach space X , closed subsets D of X and continuous right hand sides f . To ensure that solutions exist or remain in D we need an extra condition on f at the boundary points of D . Roughly speaking, this condition says that the vector f(t,x) should not point into the opposite half space determined by the tangent plane at x . It turns out that such a boundary condition and the continuity of f are sufficient for local existence in D , in case $\dim X < \infty$. If $\dim X = \infty$, we shall assume that f satisfies in addition one of the hypotheses that have been considered in the preceding chapters.

1. Boundary conditions.

Let D be a closed subset of X and $x: [0,\delta) \to D$ a solution of

$$(1) \qquad x' = f(t,x) \quad , \quad x(0) = x_0 \quad .$$

Then $\rho(x_0+hf(0,x_0)+o(h),D) = 0$ as $h \to 0+$, and therefore

$$\rho(x_0+hf(0,x_0),D) = o(h) \quad .$$

For this reason, we shall consider the condition

$$(2) \quad \lim_{\lambda \to 0+} \lambda^{-1} \rho(x+\lambda f(t,x),D) = 0 \quad \text{for } t \in J \;,\; x \in \partial D \;.$$

Obviously, (2) is satisfied at every interior point x of D . In case D is also convex, we may reformulate (2) in terms of functionals, as is shown by

Lemma 4.1. Let $D \subset X$ be closed and convex, $x \in \partial D$ and $z \in X$. Then, the following conditions for the vector z are equivalent

$$(3) \qquad \underline{\lim}_{\lambda \to 0+} \lambda^{-1} \rho(x+\lambda z,D) = 0$$

(4) If $x^* \in X^*$ and $x^*(x) = \sup_D x^*(y)$, then $x^*(z) \leq 0$.

Proof. The distance of a point $u \in X$ to a convex set D is given by

$$(5) \quad \rho(u,D) = \max\{x^*(u) - \sup_D x^*(y) : |x^*| = 1\} \quad ,$$

see e.g. [78,p.62] .
Let us assume first that (3) holds. Let $x^* \neq 0$ and

$$x^*(x) = \sup_D x^*(y) \quad ,$$

and consider $y^* = x^*/|x^*|$. By (3) and (5) ,

$$y^*(z) \leq \lambda_i^{-1} \rho(x+\lambda_i z,D) \to 0$$

for some sequence $\lambda_i \to 0+$. Hence, $x^*(z) \leq 0$. Now, we assume that (4) is true, but (3) is wrong. Then, by (5) ,

$$(6) \qquad \lambda_n^{-1}[x_n^*(x) - \sup_D x_n^*(y)] + x_n^*(z) \geq \alpha$$

for some $\alpha > 0$ and every n , where $\lambda_n \to 0+$ and $|x_n^*| = 1$. Since

$$x_n^*(x) - \sup_D x_n^*(y) \leq 0 \quad ,$$

(6) implies $x_n^*(z) \geq \alpha$ for every n and

$$x_n^*(x) - \sup_D x_n^*(y) \to 0 \quad \text{as} \quad n \to \infty \;.$$

As in the proof of Lemma 3.1 (iii) , we find

$$x_o^* \in \bigcap_{k \geq 1} \overline{\{x_n^* : n \geq k\}}^{w^*}$$

and a subsequence $(x_m^*) \subset (x_n^*)$ such that $x_m^*(z) \to x_o^*(z)$. Hence, $x_o^*(z) \geq \alpha$. Repeating the same argument at some $\bar{y} \in D$ and x , we find a subsequence (x_m^*) such that $x_m^*(\bar{y}) \to x_o^*(\bar{y})$ and $x_m^*(x) \to x_o^*(x)$. Hence,

$$x_o^*(\bar{y}) \leq \lim_{m\to\infty} \sup_D x_m^*(y) = \lim_{m\to\infty} x_m^*(x) = x_o^*(x) \quad .$$

This implies

$$x_o^*(x) = \sup_D x_o^*(y)$$

and therefore $x_o^*(z) \leq 0$, by (4) , a contradiction to $x_o^*(z) \geq \alpha > 0$.

q.e.d.

By the proof just given, it is obvious that (4) is equivalent to

$$"\lim_{\lambda\to 0+} \lambda^{-1}\rho(x+\lambda z, D) = 0" \quad ,$$

but sometimes it is easier to verify (3) . Let us consider the special case where D is a ball or a wedge, i.e. a closed convex subset K of X such that $\lambda K \subset K$ for all $\lambda \geq 0$.

Example 4.1. (i) If $D = \bar{K}_r(0)$ and $x \in \partial D$ then (3) is equivalent to "$(z,x)_+ \leq 0$" . This follows immediately from (4) and Def. 3.1 .

(ii) If D is a wedge K and $x \in \partial K$, then (3) is equivalent to

(7) If $x^* \in K^*$ and $x^*(x) = 0$, then $x^*(z) \geq 0$.

Here, K^* denotes the dual wedge

$$K^* = \{x^* \in X^* : x^*(x) \geq 0 \text{ for every } x \in K\} \quad .$$

To show this equivalence, we notice first that (4) is equivalent to

$$x^*(z) \geq 0 \text{ provided } x^*(x) = \inf_D x^*(y) \quad .$$

Since we have a wedge, $x^*(x) = \inf_K x^*(y)$ implies $x^*(x) = 0$ and $x^* \in K^*$.

2. Approximate solutions

For existence of solutions to problem (1) in a closed set, the approximate solutions given by Theorem 1.1 are of little use since all of them may be situated in the exterior of D . Since we are going to impose additional conditions on f at points of D only, we have no means to prove that a subsequence of these approximate solutions is convergent. It will turn out that the classical Euler-Cauchy polygons are more appropriate, since, by means of the boundary condition, we may modify this method in such a way that at least the corners of the polygons are in D .

Lemma 4.2. Let X be a Banach space, $D \subset X$, $J = [0,a] \subset \mathbb{R}$, $x_0 \in D$ and $D_r = D \cap \overline{K}_r(x_0)$ closed, $f: J\times D_r \to X$ continuous and $|f(t,x)| \leq c$,

(8) $\lim\limits_{\lambda\to 0+} \lambda^{-1}\rho(x+\lambda f(t,x),D) = 0$ for $t \in J$, $x \in \overline{K}_r(x_0) \cap \partial D$.

Finally, let $\varepsilon > 0$ and $b_\varepsilon = \min\{a, r/(c+\varepsilon)\}$. Then, problem (1) has an ε-approximate solution u on $[0,b_\varepsilon]$ of the following type: There is a partition $0 = t_0 < t_1 < \ldots < t_p = b_\varepsilon$ with $t_{i+1}-t_i \leq \varepsilon$ such that

(i) $u(0) = x_0$, $u(t_i) \in D_r$, u is linear in $[t_i,t_{i+1}]$, $u(t) \in \overline{K}_r(x_0)$ and $|u(t)-u(\overline{t})| \leq (c+\varepsilon)|t-\overline{t}|$ on $[0,b_\varepsilon]$.

(ii) $|u'(t)-f(t_i,u(t_i))| \leq \varepsilon$ in (t_i,t_{i+1}) .

(iii) $|f(t,x)-f(t_i,u(t_i))| \leq \varepsilon$ for $t \in [t_i,t_{i+1}]$ and $x \in \overline{K}_r(x_0)$ with $|x-u(t_i)| \leq (c+\varepsilon)(t_{i+1}-t_i)$, where we assume implicitely that f has been continuously extended to $J\times X$ in case $D_r \neq \overline{K}_r(x_0)$.

Proof. Suppose we have already defined the t_j up to t_i (for some $i \geq 0$) and the polygon u on $[0,t_i]$ such that (i) - (iii) hold. Then, we let $\delta \in (0,\varepsilon]$ be the largest number such that $t_i+\delta \leq b_\varepsilon$ and the conditions

(α) $|f(t,x) - f(t_i,u(t_i))| \leq \varepsilon$ for $t \in [t_i,t_i+\delta]$ and $x \in \overline{K}_r(x_0)$ satisfying $|x - u(t_i)| \leq (c+\varepsilon)\delta$

(β) $\rho(u(t_i) + \delta f(t_i,u(t_i)),D) \leq \frac{\varepsilon}{2}\delta$

hold simultaneously. Now, we let $t_{i+1} = t_i+\delta$, we choose $u(t_{i+1})$ as a point in D such that

$$|u(t_i) + (t_{i+1}-t_i)f(t_i,u(t_i)) - u(t_{i+1})| \leq \varepsilon(t_{i+1}-t_i)$$

which is possible by (β) , and we define

$$u(t) = \frac{u(t_{i+1})-u(t_i)}{t_{i+1}-t_i}(t-t_i) + u(t_i) \qquad \text{for} \quad t \in [t_i,t_{i+1}] \ .$$

By these definitions it is immediately seen that (i) - (iii) hold in $[0,t_{i+1}]$. Therefore, we only have to show $t_p = b_\varepsilon$ for some $p \geq 1$. Suppose on the contrary that $t_i < b_\varepsilon$ for every i , and let $t^* = \lim\limits_{i\to\infty} t_i$.

Then, in particular, the numbers $\delta = \delta_i$ determined by (α) and (β) tend to zero as $i \to \infty$. By property (i) of u, $\lim_{i\to\infty} u(t_i) = v$ exists, and $v \in D_r$ since D_r is closed. Since f is continuous, we find $\eta_o > 0$ and an index i_o such that

$$|f(t,x) - f(t_i,u(t_i))| \leq \varepsilon \quad \text{and} \quad |f(t^*,v) - f(t_i,u(t_i))| \leq \varepsilon/4$$

hold for $i \geq i_o$, $t^*-\eta_o \leq t \leq t^*$ and $x \in \overline{K}_r(x_o)$ satisfying

$$|x - u(t_i)| \leq (c+\varepsilon)\eta_o \quad .$$

Furthermore,

$$\rho(u(t_i) + \eta f(t_i,u(t_i)),D) \leq \rho(v + \eta f(t^*,v),D) + |u(t_i) - v| + \eta|f(t_i,u(t_i)) - f(t^*,v)| \quad .$$

Hence, there exist $\eta_1 \leq \eta_o$ and $i_1 \geq i_o$ such that

$$\rho(u(t_i) + \eta_1 f(t_i,u(t_i)),D) \leq \frac{\varepsilon}{2}\eta_1 \quad \text{for} \quad i \geq i_1 \quad .$$

This shows that $\delta_i \geq \eta_1$ for every $i \geq i_1$, a contradiction.

q.e.d.

The next lemma shows that if a sequence of such approximate solutions is convergent as $\varepsilon \to 0$, then its limit is a solution of problem (1).

<u>Lemma 4.3</u>. Let the hypotheses of Lemma 4.2 be satisfied and $b<\min\{a,\frac{r}{c}\}$. Consider a sequence $\varepsilon_n \to 0+$ and corresponding ε_n-approximate solutions u_n. If $\lim_{n\to\infty} u_n(t) = u(t)$ in $[0,b]$, then u is a solution of (1) in $[0,b]$.

<u>Proof</u>. Let us choose $\varepsilon_o > 0$ such that $(c+\varepsilon_o)b \leq r$ and consider $\varepsilon_n \leq \varepsilon_o$. By property (i),

$$|u_n(t) - u_n(\overline{t})| \leq (c+\varepsilon_o)|t-\overline{t}| \quad \text{for every } n \text{ and } t,\overline{t} \in [0,b] \quad .$$

Hence, (u_n) converges uniformly to u. Obviously, $u(0) = x_o$, and for $t \in (0,b]$ we have $t \in (t_{i,n},t_{i+1,n}]$ and

$$|u(t) - u_n(t_{i,n})| \leq |u(t) - u_n(t)| + (c+\varepsilon_o)\varepsilon_n \to 0 \quad \text{as} \quad n \to \infty .$$

Therefore, $u(t) \in D_r$ in $[0,b]$, since $u_n(t_{i,n}) \in D_r$ and D_r is closed. Let $\alpha_n = \max\{|u_n(t) - u(t)| : t \in [0,b]\}$. Then

$$|u(t) - x_o - \int_0^t f(s,u(s))ds| \leq \alpha_n + \int_0^t |f(s,u(s))-f(s,u_n(s))|ds +$$

$$+ 2\varepsilon_n b \to 0 \quad \text{as} \quad n \to \infty \quad .$$

Therefore, u is a solution on $[0,b]$ of (1) .

q.e.d.

3. Existence

By Lemma 4.3 , we have to look for conditions on f which guarantee convergence of the approximate solutions u_n . We start with situations where f satisfies such conditions on the whole range of every u_n .

Theorem 4.1. Let X be a Banach space, $D \subset X$, $J = [0,a] \subset \mathbb{R}$, $x_0 \in D$ and $D_r = D \cap \overline{K}_r(x_0)$ closed , $f: J \times D_r \to X$ continuous and $|f(t,x)| \le c$,

$$(8) \qquad \lim_{\lambda \to 0+} \lambda^{-1} \rho(x+\lambda f(t,x),D) = 0 \qquad \text{for} \qquad t \in J \quad , \quad x \in \overline{K}_r(x_0) \cap \partial D \quad .$$

Let $b < \min\{a,r/c\}$. Then problem (1) has a solution on $[0,b]$ with range in D_r provided one of the following extra conditions is satisfied.

(i) $\dim X < \infty$.

(ii) D_r is convex, and there is a function ω of class U_1 (cp. Theorem 3.2) such that

$$(f(t,x)-f(t,y),x-y)_- \le \omega(t,|x-y|)|x-y| \quad \text{for} \quad t \in (0,a] \text{ and } x,y \in D_r .$$

(iii) f is continuous and bounded by the constant c on $J \times \overline{K}_r(x_0)$, and the estimate in (ii) holds for $t \in (0,a]$ and every $x,y \in \overline{K}_r(x_0)$.

(iv) D_r is convex. There is a function ω as in Theorem 2.2 such that

$$\alpha(f(J \times B)) \le \omega(\alpha(B)) \qquad \text{for} \qquad B \subset D_r \quad .$$

(v) f is continuous and bounded by c on $J \times \overline{K}_r(x_0)$ and the estimate in (iv) holds for every $B \subset \overline{K}_r(x_0)$.

If (ii) or (iii) holds then the solution is unique.

Proof. Consider $\varepsilon_n \to 0$ and the ε_n-approximate solutions u_n according to Lemma 4.2 . The sequence (u_n) is equicontinuous and bounded. Hence, in case $\dim X < \infty$ there is a uniformly convergent subsequence. By Lemma 4.2 the limit is a solution in D .
Now, let us consider (ii) - (v) . In case D_r is also convex, the poly-

gons u_n lie in D_r . In general the u_n have range in $\overline{K}_r(x_o)$ only, and therefore we assume that f satisfies the estimates in the whole ball, in case (iii) or (v) holds. By (ii) of Lemma 4.2 , we have

$$|D_-'u_n(t) - f(t,u_n(t))| \leq 2\varepsilon_n \quad \text{in} \quad (0,b] \quad ,$$

where D_-' denotes the leftsided derivative. Since Proposition 2.1 and (v) of Lemma 3.2 still hold if we replace the derivative by D_-' , we may proceed as in the proofs of Theorem 2.2 and Theorem 3.2 , respectively.

q.e.d.

By means of Lemma 4.2 and Lemma 4.3 it is also easy to prove

<u>Theorem 4.2</u>. Let the hypotheses in front of (i) in Theorem 4.1 be satisfied. Let

(9) $\quad (f(t,\overline{x})-f(t,\overline{y}),x-y)_- \leq \omega(t,|x-y|)|x-y|+\omega_1(|x-\overline{x}|+|y-\overline{y}|)$

for $t \in (0,a]$, $\overline{x},\overline{y} \in D_r$ and $x,y \in \overline{K}_r(x_o)$, where ω is of class U_1 and $\omega_1: \mathbb{R}_+ \to \mathbb{R}_+$ is nondecreasing and satisfies $\omega_1(\rho) \to 0$ as $\rho \to 0$. Then problem (1) has a unique solution on $[0,b]$.

<u>Proof</u>. Let $\varepsilon_n \to 0+$; u_n and u_m be ε_n- and ε_m-approximate solutions according to Lemma 4.2 ; $\{t_o,\dots,t_p\}$ and $\{\overline{t}_o,\dots,\overline{t}_q\}$ be the corresponding partitions of $[0,b]$. Finally, let $\phi(t) = |u_n(t) - u_m(t)|$. By Lemma 4.2 and (9) , we obtain for $t \in (t_i,t_{i+1}] \cap (\overline{t}_j,\overline{t}_{j+1}]$

$$\begin{aligned}\phi(t)D^-\phi(t) &\leq (f(t_i,u_n(t_i))-f(\overline{t}_j,u_m(\overline{t}_j)),u_n(t)-u_m(t))_-+(\varepsilon_n+\varepsilon_m)\phi(t) \\ &\leq (f(t,u_n(t_i))-f(t,u_m(\overline{t}_j)),u_n(t)-u_m(t))_- + 2(\varepsilon_n+\varepsilon_m)\phi(t) \\ &\leq \omega(t,\phi(t))\phi(t) + \omega_1((c+1)(\varepsilon_n+\varepsilon_m)) + 2(\varepsilon_n+\varepsilon_m)\phi(t) \quad .\end{aligned}$$

Now, we proceed as in the proof of Theorem 3.2 .

q.e.d.

For example, a condition of type (9) is satisfied if f has one of the following properties

(i) f is Lipschitz on $(0,a]\times D_r$.

(ii) $(f(t,x)-f(t,y),x-y)_- \leq L|x-y|^2$ on $J\times D_r$ and X^* is uniformly convex (use Lemma 3.2 (v)) .

In case D_r is only known to be closed and f satisfies a condition of dissipative type on $J\times D_r$ only, it is essentially more complicated to prove existence of solutions to problem (1). We would like to estimate $|u_n(t) - u_m(t)|$ as in the proof of Theorem 3.2. In general this is impossible since we know $u_n(t) \in D$ for $t = t_0,\dots,t_p$ only. However, one can prove the following theorem which is taken from [114]; see Remark (iii).

Theorem 4.3. Let X be a Banach space, $D \subset X$, $J = [0,a] \subset \mathbb{R}$, $x_0 \in D$ and $D_r = D \cap \bar{K}_r(x_0)$ closed, $f: J\times D_r \to X$ continuous and $|f(t,x)| \leq c$,

$$(2) \quad \lim_{\lambda\to 0+} \lambda^{-1}\rho(x+\lambda f(t,x),D) = 0 \quad \text{for} \quad t\in J \;,\; x\in \bar{K}_r(x_0)\cap \partial D \;,$$

$$(10) \quad (f(t,x)-f(t,y),x-y)_+ \leq L|x-y|^2 \quad \text{for} \quad t\in J \text{ and } x,y\in D_r \;.$$

Let $b < \min\{a,r/c\}$. Then problem (1) has a unique solution on $[0,b]$.

For right hand sides satisfying a compactness condition, we have the following

Theorem 4.4. Let X be a Banach space, $D \subset X$, $J = [0,a] \subset \mathbb{R}$, $x_0 \in D$ and $D_r = D \cap \bar{K}_r(x_0)$ closed, $f: J\times D_r \to X$ continuous and $|f(t,x)| \leq c$. Let f satisfy the boundary condition (8) and

$$(11) \qquad \alpha(f(J\times B)) \leq \omega(\alpha(B)) \quad \text{for all} \quad B\subset D_r \;,$$

where $\omega: \mathbb{R}_+ \to \mathbb{R}_+$ is continuous nondecreasing and such that the initial value problem $\rho' = \omega(\rho)$, $\rho(0) = 0$ has only the trivial solution $\rho(t)\equiv 0$ on J. Then (1) has solution on $[0,b]$, where $b < \min\{a,r/c\}$.

Proof. (i) We consider a sequence (u_n) of approximate solutions, given by Lemma 4.2, such that in particular

$$|D_-^{\prime}u_n(t)-f(t,u_n(t))| \leq 1/n \quad \text{and} \quad \rho(u_n(t),D_r) \leq 1/n \quad \text{in} \quad [0,b] \;.$$

(ii) It is possible to extend f continuously to $J\times X$ by means of a formula like

$$(12) \qquad \tilde{f}(t,x) = \begin{cases} f(t,x) & \text{for } x\in D_r \;,\; t\in J \\ \sum_{\lambda\in\Lambda} \phi_\lambda(x) f(t,a_\lambda) & \text{for } x\notin D_r \;,\; t\in J \;, \end{cases}$$

where the sum is locally finite, $0 \leq \phi_\lambda(x) \leq 1$ and $\sum_{\lambda\in\Lambda} \phi_\lambda(x) \equiv 1$, $a_\lambda \in D_r$ and $|x-a_\lambda| \leq 3\rho(x,D_r)$ if $\phi_\lambda(x) \neq 0$; see [46,p.21]. We shall prove

(13) $\alpha(\tilde{f}(J\times B)) \leq \omega(\alpha(B) + 6 \sup_B \rho(x,D_r))$ for each bounded $B\subset X$.

Let B be bounded, $B = B_1 \cup B_2$ with $B_1 = D_r \cap B$ and $B_2 = B\setminus D_r$. Then (11) implies

$$\alpha(\tilde{f}(J\times B)) \leq \max\{\omega(\alpha(B)),\alpha(\tilde{f}(J\times B_2))\} \quad .$$

By (12) we have $\tilde{f}(J\times B_2) \subset \text{conv } f(J\times A)$, where

$$A = \{a_\lambda(x) : x\in B_2 \text{ and } \phi_\lambda(x) \neq 0\} \quad .$$

Therefore,

$$\alpha(\tilde{f}(J\times B_2)) = \alpha(f(J\times A)) \leq \omega(\alpha(A)) \quad .$$

Let $B_2 = \bigcup_{i=1}^p M_i$ and $A_i = \{a_\lambda(x) : x\in M_i \text{ , } \phi_\lambda(x) \neq 0\}$. Then $A = \bigcup_{i=1}^p A_i$ and

$$\text{diam } A_i \leq \text{diam } M_i + 6 \sup_{B_2} \rho(x,D_r) \text{ .}$$

This implies

$$\alpha(A) \leq \alpha(B_2) + 6 \sup_{B_2} \rho(x,D_r) \leq \alpha(B) + 6 \sup_B \rho(x,D_r) \quad ,$$

and therefore (13) is true.

(iii) Let $B_k(t) = \{u_n(t) : n \geq k\}$ and $\phi_k(t) = \alpha(B_k(t))$. As in the proof of Theorem 2.2 , we obtain by means of (13)

$$D^-\phi_k(t) \leq \omega(\phi_k(t) + 6 \sup_{B_k(t)} \rho(x,D_r)) + \frac{2}{k} \quad \text{in} \quad (0,b] \text{ .}$$

By step (i) , we have $\sup\{\rho(x,D_r) : x\in B_k(t)\} \leq 1/k$. Let $\psi_k(t)=\phi_k(t)+\frac{6}{k}$. Then $D^-\psi_k(t) \leq \omega(\psi_k(t)) + \frac{2}{k}$ in $(0,b]$, $\psi_k(0) = 6/k$. This implies $\psi_k(t) \to 0$ and therefore $\phi_k(t) \to 0$. Hence,

$$\alpha(\{u_n(t) : n \geq 1\}) = \phi_k(t) \to 0 \quad \text{for every} \quad t\in[0,b] \quad .$$

Therefore (u_n) has a uniformly convergent subsequence. Let u be the limit. Then u is a solution of (1) with $\tilde{f}$ instead of f , by Lemma 4.3 . But

$$\rho(u(t),D_r) = \lim_{n\to\infty} \rho(u_n(t),D_r) = 0 \quad .$$

Therefore, u is a solution of (1) .

q.e.d.

By means of these local theorems, global existence may be proved along the lines of sec. 3.4 . The explicite formulation is left to the reader.

4. Examples

In this section we want to illustrate by means of simple examples how the existence results of this chapter may be applied.

(i) Branching processes. Let us consider the countable system

$$x_i' = -a_{ii}x_i + \sum_{j\neq i} a_{ij}x_j \quad , \qquad x_i(0) = c_i \quad , \tag{14}$$

where $c_i \geq 0$ and $\sum_{i\geq 1} c_i = 1$, $a_{ij} \geq 0$ and $\sum_{i\neq j} a_{ij} = a_{jj}$; see the Introduction. Since the x_i should be probabilities we consider $X = l^1$ and we look for solutions in the standard cone

$$K = \{x \in l^1 : x_i \geq 0 \text{ for all } i \geq 1\}$$

such that $|x(t)| \equiv 1$.

Suppose that the matrix A corresponding to system (14) defines a bounded operator from l^1 into l^1 . By the properties of (a_{ij}) this assumption is equivalent to $\sup_i a_{ii} < \infty$. If we find a solution in K then $|x(t)| \equiv 1$ is automatically satisfied, since

$$\frac{d}{dt}|x(t)| = \sum_{i\geq 1} x_i'(t) = -\sum_{i\geq 1} a_{ii}x_i + \sum_{j\geq 1}\Big(\sum_{i\neq j} a_{ij}\Big)x_j = 0$$

and

$$\sum_{i\geq 1} x_i(0) = \sum_{i\geq 1} c_i = 1 \quad .$$

Now, the existence of a unique (global) solution is a simple consequence of Theorem 4.1 . In fact, $f(t,x) = Ax$ is Lipschitz and condition (8) is satisfied since, by Example 4.1 , $x \in \partial K$ and

$$z \in K^* = \{z \in l^\infty : z_i \geq 0 \text{ for all } i \geq 1\}$$

and $\sum_{i\geq 1} z_i x_i = 0$ imply

$$\sum_{i\geq 1} z_i(Ax)_i = \sum_{i\geq 1} z_i\Big(-a_{ii}x_i + \sum_{j\neq i} a_{ij}x_j\Big) = \sum_{i\geq 1} z_i\Big(\sum_{j\neq i} a_{ij}x_j\Big) \geq 0.$$

(ii) Boundary points with outer cone condition. Let X be a Banach space and $D \subset X$. Let us say that $x_o \in \partial D$ satisfies an outer cone condition if there exists a closed convex set $C \subset X$ such that $\overset{\circ}{C} \neq \emptyset$ and $D \cap C = \{x_o\}$. In case D is convex, this condition is satisfied if and only if there is a supporting hyperplane at x_o , by Mazur's theorem.

Theorem 4.5. If $D \subset X$ is closed and

$$M = \{x \in \partial D : x \text{ satisfies an outer cone condition}\}$$

then M is dense in ∂D .

Proof. Suppose M is not dense in ∂D . Then there exists $x_o \in \partial D$ and $r > 0$ such that $M \cap \bar{K}_r(x_o) = \emptyset$. We choose $y_o \in K_r(x_o) \setminus D$ and we let $f(x) \equiv y_o - x_o$ for $x \in \bar{K}_r(x_o)$. Obviously, $x(t) = x_o + t(y_o - x_o)$ is the unique solution of $x' = f(x)$, $x(0) = x_o$. Let us apply Theorem 4.1 . Clearly, condition (v) with $\omega \equiv 0$ and $c < r$ is satisfied. In order to check the boundary condition (8) , let $z = y_o - x_o$ and $x \in \bar{K}_r(x_o) \cap \partial D$, $\varepsilon > 0$ and $\delta > 0$, and consider

$$C = \{x + \lambda z + \varepsilon\lambda v : 0 \leq \lambda \leq \delta , |v| = 1\} .$$

C is closed convex with $\overset{\circ}{C} \neq \emptyset$ and $x \in C$. Since $x \notin M$, we have

$$x + \lambda z + \varepsilon\lambda v \in D \quad \text{for some } \lambda \in (0,\delta] \text{ and some } v \text{ with } |v| = 1 .$$

Hence, $\rho(x+\lambda f(x),D) \leq \varepsilon\lambda$ for this λ . Since ε and δ have been arbitrary, this implies (8) . Therefore, $x(t) \in D$ in $[0,1]$, in particular $x(1) = y_o \in D$, a contradiction.

q.e.d.

(iii) Fixed points of nonexpansive maps. Let us start with a simple extension of Banach's fixed point theorem.

Theorem 4.6. Let X be a Banach space, $D \subset X$ closed, $T: D \to X$ Lipschitz with constant $L < 1$,

$$(15) \qquad \lim_{\lambda \to 0+} \lambda^{-1} \rho(x+\lambda(Tx-x),D) = 0 \quad \text{for all} \quad x \in \partial D .$$

Then T has a unique fixed point.

Proof. Consider $u' = Tu - u$, $u(0) = x \in D$. By Theorem 4.4 and Theorem 3.3 this problem has a unique solution $u(\cdot,x)$ on $[0,\infty)$. For some fixed $p > 0$, let $Ux = u(p,x)$. Then $U: D \to D$ and $|Ux-Uy| \leq |x-y|\exp(-(1-L)p)$; cp. the proof to Theorem 3.6 . Therefore, U has a unique fixed point x_o . This implies $u(t,x_o) \equiv x_o$, i.e. $Tx_o = x_o$.

q.e.d.

For example, (15) is obviously satisfied if D is also convex and $T(\partial D) \subset D$. For nonexpansive maps we may prove fixed point theorems along the lines of sect. 3.6 . For instance

<u>Theorem 4.7.</u> Let X be a uniformly convex Banach space, $D \subset X$ closed and convex, $T: D \to X$ nonexpansive, and the boundary condition (15) be satisfied. Suppose also that either $|Tx-x| \to \infty$ as $|x| \to \infty$ or $(Tx,x)_+ \le |x|^2$ for all $x \in D$ with $|x| = r$, for some $r > 0$ such that $D \cap \bar{K}_r(0) \neq \emptyset$. Then T has a fixed point.

<u>Proof.</u> Suppose first that $|Tx-x| \to \infty$ as $|x| \to \infty$. We may assume $0 \in D$, considering $D-x_0$ and $T(x+x_0)-x_0$ instead of D and T, for some fixed $x_0 \in D$. Let us consider $T_n = T - \frac{1}{n}I$. The mapping $T_n - I$ also satisfies the boundary condition (8), since $x^* \in X^*$ and $x \in \partial D$ and

$$x^*(x) = \sup_D x^*(y)$$

imply

$$x^*(T_n x-x) = x^*(Tx-x) - \frac{1}{n}x^*(x) \le - \frac{1}{n}x^*(x) \le 0 \quad .$$

As in the proof of Theorem 3.7 (i), we therefore find a bounded sequence $(x_n) \subset D$ such that $x_n-Tx_n \to 0$ as $n \to \infty$. Since X is uniformly convex and T is nonexpansive, $I-T$ maps bounded closed convex sets onto closed sets ; see [21,Theorem 8.4] . Therefore, there exists an $x \in \overline{\text{conv}\{x_n : n \ge 1\}} \subset D$ with $Tx = x$.
Now, let us assume that $D_r = D \cap \bar{K}_r(0) \neq \emptyset$ and $(Tx,x)_+ \le |x|^2$ for $|x|=r$. Then $(Tx-x,x)_+ \le 0$ for $x \in D$ with $|x| = r$. Therefore, the solutions of $u' = Tu-u$ starting in the closed bounded convex set D_r remain in D_r (see Example 4.1) and we can apply Lemma 3.3 to $\{U(t) : t \ge 0\}$, where $U(t)x = u(t,x)$ for $x \in D_r$.

q.e.d.

In the proof of another fixed point theorem we need the following simple

<u>Proposition 4.1.</u> Let X be a Banach space, $u: \mathbb{R} \to X$ p-periodic,

$$|u(t)-u(s)| \le L|t-s| \quad \text{and} \quad \int_0^p u(s)ds = 0 \quad .$$

Then $p \ge \frac{4}{L} \max_{\mathbb{R}} |u(t)|$.

<u>Proof.</u> For fixed t we integrate the identity $u(t) = u(s)+(u(t)-u(s))$ over $[t-p/2,t+p/2]$ to obtain

$$p|u(t)| \leq |\int_0^p u(s)ds| + L\int_{t-p/2}^{t+p/2} |t-s|ds = \frac{L}{4}p^2$$

q.e.d.

Theorem 4.8. Let X be a Banach space, $D \subset X$ closed bounded and convex, $T: D \to X$ continuous and

(16) $$\alpha(T(B)) \leq k\alpha(B) \quad \text{for some } k < 1 \text{ and all } B \subset D \ .$$

Let T satisfy the boundary condition (15) and suppose that the problem

(17) $$u' = Tu-u \quad , \quad u(0) = x$$

has at most one solution, for each $x \in D$. Then T has a fixed point.

Proof. Since $T-I$ satisfies (11) with $\omega(\rho) = (k+1)\rho$, problem (17) has a solution $u(t,x)$ on $[0,\infty)$. It is unique by assumption. Let

$$U(t)x = u(t,x) \quad \text{for} \quad t \geq 0 \quad \text{and} \quad x \in D \ .$$

Then $U(t): D \to D$ is continuous. Let $B \subset D$ and $\phi(t) = \alpha(e^tU(t)B)$. By (17) , $(e^tu)' = e^tTu$ and therefore (cp. the proof to Theorem 2.2)

$$\begin{aligned} D^-\phi(t) &\leq \lim_{\tau\to 0+} \alpha(\bigcup_{J_\tau} e^sTU(s)B) \\ &\leq \lim_{\tau\to 0+} [\alpha(e^tT(\bigcup_{J_\tau} U(s)B)) + 2e^t(1-e^{-\tau})c] \\ &\leq ke^t \lim_{\tau\to 0+} \alpha(\bigcup_{J_\tau} U(s)B) = k\phi(t) \quad , \end{aligned}$$

where $c = \sup\{|Tx| : x \in D\}$. Since $\phi(0) = \alpha(B)$, this implies

$$\phi(t) \leq e^{kt}\alpha(B)$$

and therefore

$$\alpha(U(t)B) \leq e^{-(1-k)t}\alpha(B) \quad \text{for } B \subset D \ , \ t \geq 0 \ .$$

Hence, we can apply Lemma 2.3 to obtain a fixed point x_p of $U(p)$, for every $p > 0$. This implies that $u(\cdot,x_p)$ is p-periodic. Now, consider a sequence $p_n \to 0$. Let

$$u_n = u(\cdot,x_{p_n}) \quad \text{and} \quad z_n = \frac{1}{p_n}\int_0^{p_n} u_n(s)ds \quad ,$$

$$v_n(t) = \begin{cases} u_n(t) - z_n & \text{for } t \geq 0 \\ v_n(t+(k+1)p_n) & \text{for } t \in (-(k+1)p_n,-kp_n] \ , \ k = 0,1,2,\ldots \ . \end{cases}$$

Obviously, v_n is p_n-periodic , $|v_n(t) - v_n(s)| \leq L|t-s|$ with

$$L = \sup\{|Tx-x| : x\in D\} \quad \text{and} \quad \int_0^{p_n} v_n(s)ds = 0 \quad .$$

Therefore, Proposition 4.1 implies

$$\max_{\mathbb{R}} |v_n(t)| \to 0 \quad \text{as} \quad n \to \infty \quad .$$

Let $B = \{z_n : n \geq 1\}$ and $C = \{x_{p_n} : n \geq 1\}$. Since $v_n(t) \to 0$ as $n \to \infty$, we have

$$\alpha(B) = \alpha(U(t)C) \leq e^{-(1-k)t}\alpha(C) \to 0 \quad \text{as} \quad t \to \infty \quad .$$

Hence, we may assume $z_n \to z$ for some $z \in X$. This implies $u_n(t) \to z$ uniformly in $t \geq 0$ and therefore $u(t) \equiv z$ is a solution of $u' = Tu-u$, $u(0) = z$. Thus, $Tz = z$.

q.e.d.

5. Remarks

(i) In finite dimensional spaces, a thorough discussion of flow invariance by means of condition (8) , as well as applications to stability reachable sets in control problems and properties of Peano's funnel may be found in Yorke [194] , [195] . In case $\dim X < \infty$, Theorem 4.1 has been proved by Nagumo [128] and has been rediscovered several times ; see Brezis [18] , Crandall [38] and Hartman [72] . Further remarks will be given in § 5 .

(ii) We have noticed already that we may replace $\underline{\lim}$ by lim in Lemma 4.1. In Lemma 4.2 we have established the existence of certain ε-approximate solutions to (1) by means of (8) . Hereafter, it is easy to show that $\lim_{\lambda\to 0+} \lambda^{-1}\rho(x+\lambda f(t,x),D)$ exists and that this convergence is even uniform with respect to x from compact subsets of $K_r(x_0) \cap \partial D$; see Martin [115] for details.

(iii) Lemma 4.2 , Lemma 4.3 and Theorem 4.1 are essentially taken from Martin [114] , [115] . For convex D , related results have been proved before by Crandall [37] . Conditions like (9) in Theorem 4.2 have also been considered in these papers.
The difficulty in the proof of Theorem 4.3 consists in the fact that $u_n(t_i)$ is in D but $u_m(t_i)$ may be outside of D , and therefore (10) is of no use in estimating $|u_n(t) - u_m(t)|$. For this reason Martin [114]

considers piecewise linear functions v_n and v_m which are close to u_n and u_m and have values in D at all points $t \in \{t_o,\dots,t_p\} \cup \{\bar{t}_o,\dots,\bar{t}_q\}$. The construction also depends on the fact that $(\cdot,\cdot)_+$ is upper semi-continuous ; see Lemma 3.2 .
Lakshmikantham/A.R.Mitchell/R.W.Mitchell [100] have replaced (10) by the more general condition (5) from Remark (iii) to § 3 , with $\omega(t,\rho)$ continuous and increasing in ρ .
In a global result corresponding to Theorem 4.3 , Martin [114] assumes that D is closed, L in (10) may be a continuous function on $[0,\infty)$ and f satisfies in addition one of the following conditions

(H1) f maps bounded sets into bounded sets.

(H2) On bounded subintervals of $[0,\infty)$, f is uniformly continuous in t , uniformly with respect to x from bounded subsets of D .

In case D = X , Lovelady/Martin [109] have proved this result without (H1) or (H2) , by means of semigroup theory. It seems to be an open question wether (H1) and (H2) are also redundant for general closed D .

(iv) Theorem 4.4 is taken essentially from Volkmann [186] . It has been proved before by Martin [115] under the additional assumption that f is uniformly continuous.

(v) Theorem 4.5 is taken from Browder [23] , where it plays an essential role in the study of the Fredhölm alternative for certain nonlinear maps. The simple proof presented here is taken from Volkmann [185]. See also Phelps [140] for a survey of such results.

(vi) A particular case of Theorem 4.2 has been used by Bourguignon/Brezis [17] to study the motion of an incompressible perfect fluid which is described as follows.
Given a bounded domain $\Omega \subset R^n$ (with smooth boundary and outward normal ν) , an external force $f(x,t)$ and an initial velocity $u_o(x)$ on Ω , the velocity u and the pressure p satisfy the Euler equation

$$\frac{\partial u_i}{\partial t} + \sum_{j=1}^{n} u_j \frac{\partial u_i}{\partial x_j} = f_i + \frac{\partial p}{\partial x_i} \quad \text{on } \Omega\times(0,T) \quad , \quad i = 1,\dots,n \quad ,$$

and the side conditions div u = 0 on $\Omega\times(0,T)$, $(u,\nu) = 0$ on $\partial\Omega\times(0,T)$ and $u(x,0) = u_o(x)$ on Ω .
Since u is divergence free, it is easy to eliminate the pressure. Then, the configuration $\eta(x,t)$ of the fluid given by

$$\frac{\partial \eta}{\partial t} = u(\eta(x,t),t) \quad \text{and} \quad \eta(x,0) = x \quad ,$$

is introduced as the new unknown function. $\eta(\cdot,t)$ is regarded as an element of some Sobolev space $W^{s,p}(\Omega;R^n)$, since the given data f , u_0 are assumed to be in some related Sobolev spaces. By means of the Euler equation, one is then led to an initial value problem $w' = F(t,w)$, $w(0) = w_0$ for the pair $w = (\eta, \frac{\partial\eta}{\partial t})$ in the Banach space

$$X = W^{s,p}(\Omega;R^n)\times W^{s,p}(\Omega;R^n) \quad .$$

Actually, F is defined and locally Lipschitz on some closed subset D of X only, since the functions η have to be certain diffeomorphisms in order to establish the transformation and to have the equivalence of the new problem with the old one for u . Finally, since there is no flux through the surface, i.e. $(u,\nu) = 0$, one can show that F satisfies the boundary condition (2) .

(vii) Results related to Theorem 4.6 and Theorem 4.7 may be found e.g. in Browder [21] , Crandall [37] , Deimling [49] , Martin [114] and Vidossich [178] . Bounds for periods, more general than Proposition 4.1, may be found in Lasota/Yorke [102] and Vidossich [177] .
In case $\overset{\circ}{D} \neq \emptyset$, Theorem 4.8 can be proved by means of degree theory for k-set contractions without the hypothesis that (17) has at most one solution, even under the more general boundary condition of Leray/ Schauder, i.e.

"If $Tx = \lambda x$ for some $x \in \partial D$ then $\lambda \leq 1$" (in case $0 \in D$) ;

see Nussbaum [132]. In the special case $T(\partial D) \subset D$ we also do not need the hypothesis on (17) . To see this, we notice first that we may approximate T by locally Lipschitz maps T_n such that $T_n(\partial D) \subset D$ (choose $a_\lambda \in \partial D$ if $V_\lambda \cap \partial D \neq \emptyset$ in the proof of Lemma 1.1) and T_n satisfies (16). Then Theorem 4.8 gives us a fixed point $x_n = T_n x_n$, and it is easy to see that a subsequence of (x_n) converges to the fixed point of T . In case D is only closed we do not know whether Theorem 4.8 is true without the hypothesis on (17) .
In case T is locally Lipschitz, Theorem 4.8 holds if T satisfies only "$\alpha(T(B)) < \alpha(B)$ for $\alpha(B) \neq 0$" instead of (16) , i.e. if T is only "condensing". To see this, consider $T_n = k_n T$ with $k_n \in (0,1)$ and $k_n \to 1$ as $n \to \infty$, and assume without loss of generality that $0 \in D$. The map T_n satisfies (15) since $x \in \partial D$ and $x^*(x) = \sup_D x^*(y)$ imply $x^*(x) \geq 0$ and $x^*(Tx-x) \leq 0$, whence $x^*(k_n Tx) \leq x^*(k_n x) \leq x^*(x)$, and therefore $x^*(T_n x-x) \leq 0$. Since T_n is also locally Lipschitz and (16) is satisfied with $k = k_n$, we find a fixed point $x_n = T_n x_n$. Now, it is easy to see $\alpha(B) \leq \alpha(T(B))$ for $B = \{x_n : n \geq 1\}$, and therefore $\alpha(B) = 0$. This implies that T has a fixed point.

§ 5 Flow invariance and differential inequalities

In case $X = \mathbb{R}^1$ it is well known that the inequalities

$$u' - f(t,u) < v' - f(t,v) \text{ in } (0,a) \text{ and } u(0) < v(0) \quad ,$$

for functions u , v which are continuous in $[0,a)$ and differentiable in $(0,a)$, imply $u(t) < v(t)$ in $[0,a)$, without any condition on f . It is also well known that this comparison theorem holds for $X = \mathbb{R}^n$ provided f has a certain monotonicity property called "quasimonotonicity" , but again it is not necessary to assume any regularity of f like continuity, for instance ; see [97] , [187] . In other words, comparison theorems may be established without knowledge of existence theorems for the corresponding initial value problem.
This point of view may also be taken in the problem of flow invariance. In order that a set D be (forward) invariant with respect to f , we have to show that $x(0) \in D$ and $x' = f(t,x)$ in $(0,a)$ imply $x(t) \in D$ in $[0,a)$. Hence, we need not worry about existence, but we have to look for properties of f at ∂D which do not allow the solutions to leave the set D .
Evidently, the first problem is related to the second one, since given $v(0) - u(0)$ in the interior of the standard cone

$$K = \{x \in \mathbb{R}^n : x_i \geq 0 \text{ for } i = 1,\dots,n\} \quad ,$$

we look for properties of f such that $v(t) - u(t)$ remains in the interior of K . In the sequel, we shall investigate these questions in an arbitrary real normed linear space X .

1. Boundary conditions.

Let $D \subset X$, $x \in \partial D$ and $z \in X$. In § 4 we have considered the condition $\lim\limits_{\lambda \to 0+} \lambda^{-1} \rho(x+\lambda z, D) = 0$. In case X is an inner product space and $D = \overline{K}_r(0)$, this condition is equivalent to $(z,x) \leq 0$; see Example 4.1 . We notice that in this example the vector x points in the direction of the outer normal. To extend this concept to more general sets D , we introduce

Definition 5.1. Let X be a normed linear space (NLS) , $D \subset X$ and $x \in \partial D$. A vector $\nu \in X$ is said to be an outer normal to D at x , if $\nu \neq 0$ and $K_{|\nu|}(x+\nu) \cap D = \emptyset$.

In case D is a closed convex set the existence of an outer normal ν at $x \in \partial D$ implies the existence of a supporting hyperplane at x . Therefore, outer normals need not exist. Actually, since we have not assumed that X is complete, there are closed sets without supporting hyperplane at any boundary point ; see [86] .
Now, let N(x) denote the set of all outer normals at $x \in \partial D$ and consider the condition $(z,\nu)_- \leq 0$ to be vacuously fulfilled if $N(x) = \emptyset$.

Proposition 5.1. Let X be a real NLS , $D \subset X$ closed, $x \in \partial D$ and $z \in X$. Then $\underline{\lim}_{\lambda \to 0+} \lambda^{-1}\rho(x+\lambda z, D) = 0$ implies $(z,\nu)_- \leq 0$, for every $\nu \in N(x)$.

Proof. Let $\nu \in N(x)$. At first, we show that $x+\lambda z \in K_{|\nu|}(x+\nu)$ is impossible for $\lambda > 0$. Suppose on the contrary that

$$\phi(\lambda) = |x+\lambda z-x-\nu| = |\lambda z-\nu| < |\nu| \quad \text{for some } \lambda = \lambda_o > 0 \quad .$$

Since ϕ is a convex function,

$$\phi(\lambda) \leq \lambda\lambda_o^{-1}\phi(\lambda_o) + (1-\lambda\lambda_o^{-1})|\nu| \quad \text{for} \quad 0 < \lambda \leq \lambda_o \quad ,$$

and therefore

$$\lambda^{-1}\rho(x+\lambda z,D) \geq \lambda_o^{-1}(|\nu|-\phi(\lambda_o)) \quad \text{for} \quad 0 < \lambda \leq \lambda_o \quad ,$$

a contradiction. Hence, we have $|\nu-\lambda z| \geq |\nu|$ for every $\lambda \geq 0$. Let us show that this relation implies the existence of an $x^* \in F\nu$ such that $x^*(z) \leq 0$.
Consider a sequence $\lambda_n \to 0+$, let $x_n^* \in F(\nu-\lambda_n z)$ and $y_n^* = |x_n^*|^{-1}x_n^*$. Then

$$|\nu| \leq |\nu-\lambda_n z| = \langle \nu-\lambda_n z, y_n^* \rangle \leq |\nu| - \lambda_n\langle z, y_n^* \rangle \quad .$$

Hence, $\langle z,y_n^* \rangle \leq 0$ and $\langle \nu,y_n^* \rangle \to |\nu|$ as $n \to \infty$. Since $|y_n^*| = 1$ for every n , we find $y^* \in X^*$ such that $|y^*| \leq 1$, $y^*(\nu) = |\nu|$ and $y^*(z) \leq 0$; cp. the proof of Lemma 3.1 (iii) . Therefore, $y^*(\nu) = |\nu|$ and $|y^*| = 1$. Hence, $x^* = |\nu|y^*$ is in $F\nu$ and $x^*(z) \leq 0$ is satisfied.

q.e.d.

The following simple example shows that we can not replace $(z,\nu)_-$ by $(z,\nu)_+$ in Proposition 5.1 .

Example 5.1. Let $X = \mathbb{R}^2$ with norm $|x| = |x_1| + |x_2|$, $D = \{x : x_2 \leq 0\}$, $x = 0$ and $z = (1,0)$. Obviously, $\rho(0+\lambda z,D) = 0$ for $\lambda > 0$, and $\nu = (0,1)$ is an outer normal to D at $x = 0$. But by Example 3.1 (iii) we have $(z,\nu)_+ = 1$ and $(z,\nu)_- = -1$.

2. Flow invariance

Let us recall that a set $D \subset X$ is said to be a distance set if to each $x \in X$ there exists a $y \in D$ such that $\rho(x,D) = |x-y|$. Evidently , a distance set is closed. We also need the following

Proposition 5.2. Let X be a real NLS . Then

$$(y,x)_+ = |x| \lim_{\lambda\to 0+} \frac{|x+\lambda y| - |x|}{\lambda} \quad \text{for all } x,y \in X \ .$$

Proof. Since $\lambda \to |x+\lambda y|$ is convex, it is easy to see that

$$\lambda \quad\to\quad \lambda^{-1}(|x+\lambda y| - |x|)$$

is monotone increasing in $\lambda > 0$, and since this function is also bounded from below,

$$\phi(y) \quad = \quad \lim_{\lambda\to 0+} \lambda^{-1}(|x+\lambda y| - |x|)$$

exists. Furthermore,

$$(1) \qquad \phi(y_1+y_2) \leq \phi(y_1) + \phi(y_2) \quad \text{and} \quad \phi(ty) = t\phi(y) \text{ for } t > 0.$$

By definition of $(y,x)_+$, it follows immediately that

$$-\phi(-y)|x| \leq (y,x)_+ \leq \phi(y)|x| \quad .$$

Now, let $X_o = \{tx+sy : t,s \in \mathbb{R}\}$ and define the linear functional $f: X_o \to \mathbb{R}$ as $f(tx+sy) = t|x| + s\phi(y)$. Since $\phi(tx+sy) = t|x| + \phi(sy)$ and since $\phi(y) \geq -\phi(-y)$ implies $s\phi(y) \leq \phi(sy)$, we have $f(z) \leq \phi(z)$ for all $z \in X_o$. By the properties (1) , we may apply the Hahn/Banach-Theorem to obtain $x_o^* \in X^*$ which is an extension of f and satisfies $x_o^*(z) \leq \phi(z)$ for all $z \in X$. Therefore, $x_o^*(y) = \phi(y)$ and $x_o^*(x) = |x|$, and $|x_o^*| = 1$ since $x_o^*(z) \leq \phi(z) \leq |z|$ for all $z \in X$. Hence, $x^* = |x|x_o^* \in Fx$ and $x^*(y) = \phi(y)|x|$.

q.e.d.

Theorem 5.1. Let X be a real NLS , $\Omega \subset X$ an open set and $D \subset X$ a distance set with $D \cap \Omega \neq \emptyset$. Let $f\colon (0,a)\times\Omega \to X$ be such that

(i) $$(f(t,x) - f(t,y),x-y)_+ \leq \omega(t, |x-y|)|x-y|$$

$$\text{for } x \in \Omega\setminus D\ ,\ y \in \Omega\cap\partial D\ ,\ t \in (0,a)\ ,$$

where $\omega\colon (0,a)\times\mathbb{R}^+ \to \mathbb{R}$ is such that $\rho(t) \leq 0$ in $(0,\tau) \subset (0,a)$ whenever $\rho\colon [0,\tau) \to \mathbb{R}^+$ is continuous, $\rho(0) = 0$ and $D^+\rho(t) \leq \omega(t,\rho(t))$ for every $t \in (0,\tau)$ with $\rho(t) > 0$.

(ii) If $x \in \Omega\cap\partial D$ is such that $N(x) \neq \emptyset$ then either

$$\lim_{\lambda\to 0+} \lambda^{-1}\rho(x+\lambda f(t,x),D) = 0 \quad\text{or}\quad (f(t,x),\nu)_+ \leq 0$$

for all $\nu \in N(x)$ and $t \in (0,a)$.

Then $D\cap\Omega$ is forward invariant with respect to f , i.e. any continuous $x\colon [0,b) \to \Omega$, such that $x(0) \in D$ and $x' = f(t,x)$ in $(0,b)$, satisfies $x(t) \in D$ in $[0,b)$.

Proof. Let $\rho(t) = \rho(x(t),D)$ and suppose that $\rho(t) > 0$ for some $t > 0$. Since D is a distance set, we find $p \in \Omega\cap\partial D$ such that $\rho(t) = |x(t)-p|$. Hence, $\nu = x(t)-p \in N(p)$. Suppose first that

$$(f(t,p),\nu)_+ \leq 0$$

and consider

$$\phi(s) = |x(t+s)-p| = |x(t)-p+sf(t,x(t))| + o(s)$$

for $s \to 0+$.
By Proposition 5.2 , we have

$$\begin{aligned}\phi(0)D^+\phi(0) &= (f(t,x(t)),\nu)_+\\ &\leq (f(t,x(t))-f(t,p),x(t)-p)_+ + (f(t,p),\nu)_+\\ &\leq \omega(t,\phi(0))\phi(0) \quad .\end{aligned}$$

Since $\phi(0) = \rho(t)$ and $D^+\phi(0) \geq D^+\rho(t)$, this implies

$$D^+\rho(t) \leq \omega(t,\rho(t)) \quad .$$

In case $\rho(p+\lambda f(t,p),D) = o(\lambda)$ as $\lambda \to 0+$, a simple calculation yields

$$\rho(t+s) \leq |x(t) - p + s(f(t,x(t)) - f(t,p)) \quad +$$

$$+ \rho(p + sf(t,p),D) + o(s) \quad \text{as} \quad s \to 0+ \quad ,$$

and therefore Proposition 5.2 and (i) imply again

$$D^{+}\rho(t) \leq \omega(t,\rho(t)) \ .$$

Hence, (i) implies $\rho(t) \equiv 0$ in $[0,b)$. q.e.d.

In case dim $X < \infty$ and D is closed, D is a distance set. If, however, dim $X = \infty$ then the assumption that D be a distance set is rather restrictive. Clearly, it is satisfied if D is either compact or a closed convex subset of a reflexive X .

Theorem 5.2. Let X be a real NLS , $\Omega \subset X$ open and $D \subset X$ closed with $D \cap \Omega \neq \emptyset$. Suppose f: $(0,a)\times\Omega \to X$ satisfies

$$\lim_{\lambda\to 0+} \lambda^{-1}\rho(x+\lambda f(t,x),D) = 0 \quad \text{for} \quad x \in \Omega \cap \partial D \quad \text{and} \quad t \in (0,a) \quad .$$

Then $D \cap \Omega$ is forward invariant with respect to f if one of the following hypotheses is true.

(i) D is closed convex with $\overset{\circ}{D} \neq \emptyset$ and f is locally Lipschitz.

(ii) X is complete and

$$|f(t,x) - f(t,y)| \leq \omega(t,|x-y|) \quad \text{for } t \in (0,a) \ , \ x \in \Omega \setminus D \ \text{and } y \in \Omega \cap \partial D \quad ,$$

where ω is as in Theorem 5.1 and such that $\overline{\lim}_{h\to 0+} \omega(t,\rho+h) \leq \omega(t,\rho)$ for $t \in (0,a)$ and $0 < \rho < \infty$.

Since a proof of this theorem is lengthy , we refer to Remark (ii) .

The following example shows that $\overset{\circ}{D} \neq \emptyset$ in (i) and the completeness of X in (ii) of Theorem 5.2 can not be omitted.

Example 5.2. Let X be the space of real polynomials restricted to $J = [0,1]$ endowed with the L^2-norm. Let $D = \{x \in X : x(s) \geq 0 \text{ in } J\}$ and f: $X \to X$ be defined by

$$f(x)(s) = s \int_0^1 (12\tau-6)x(\tau)d\tau \quad .$$

Clearly, f is Lipschitz in X and D is a wedge with $\overset{\circ}{D} = \emptyset$. By the representation theorem of Riesz, $x^{*} \in D^{*}$ is given by

$$x^*(x) = \int_0^1 x(s)\phi(s)ds$$

for some function $\phi \in L^2(J)$ with $\phi(s) \geq 0$ a.e. By Example 4.1 , f satisfies the boundary condition provided $x^* \in D^*$ and $x^*(x) = 0$ imply $x^*(f(x)) \geq 0$. But $x^*(x) = 0$ implies either $x(s) \equiv 0$ or $\phi(s) = 0$ a.e. Therefore $x^*(f(x)) \geq 0$.
Now, let $v: J \to X$ be defined by $v(t)(s) = 1-e^t s$. Then $v(0) \in D$ and $v'(t) = f(v(t))$ in J , but $v(t) \notin D$ for $t > 0$.

3. Differential inequalities

Let us recall that a function $f: \mathbb{R}^n \to \mathbb{R}^n$ is said to be quasimonotone if it is monotone increasing in its components outside the leading diagonal , i.e. $x \leq y$ and $x_i = y_i$ imply $f_i(x) \leq f_i(y)$. It is easy to see that f is quasimonotone if and only if

(2) $(f(y) - f(x),z) \geq 0$ whenever $x \leq y$, $z \geq 0$ and $(y-x,z) = 0$.

In order to extend this concept to an arbitrary real NLS , we first have to take care for "$\leq$" . To this end we consider a wedge $K \subset X$, and we define $x \leq y$ be equivalent with $y-x \in K$. Clearly, $\leq$ depends on the choice of K . In case $\mathring{K} \neq \emptyset$ we define $x < y$ to be equivalent with $y-x \in \mathring{K}$. It is easy to see that "$\leq$" and "$<$" have many of the usual properties.
The following proposition will be useful in the sequel

Proposition 5.3. Let X be a real NLS , $K \subset X$ a wedge with $\mathring{K} \neq \emptyset$, and $x_o \in \partial K$. Then there exists an $x^* \in \mathring{K}^* = \{x^* \in K : x^*(x) > 0 \text{ for all } x \in \mathring{K}\}$ such that $x^*(x_o) = 0$.

Proof. By Mazur's separation theorem for convex sets, there exists $y^* \in X^*$ such that

$$\sup_K y^*(x) = y^*(x_o) \quad \text{and} \quad y^*(x) < y^*(x_o) \quad \text{for all } x \in \mathring{K} .$$

Since $\lambda K \subset K$ for every $\lambda \geq 0$, this implies $y^*(x_o) = 0$. Hence, $x^* = -y^* \in \mathring{K}^*$ and $x^*(x_o) = 0$.

q.e.d.

As a simple consequence let us mention that $x > 0$ and $y \geq 0$ imply $x+y>0$.

The natural extension of quasimonotonicity to arbitrary real NLS is now given by

Definition 5.2. Let X be a real NLS , $K \subset X$ a wedge and "$\leq$" be defined by K . Given $D \subset X$ and $f: D \to X$, f is said to be quasimonotone (with respect to K) if

$$x \leq y \ , \ x^* \in K^* \quad \text{and} \quad x^*(x-y) = 0 \quad \text{imply} \quad x^*(f(x) - f(y)) \leq 0 \ .$$

In case f depends also on $t \in \mathbb{R}$, f is called quasimonotone if $f(t,\cdot)$ is quasimonotone, for every t under consideration.

Evidently, this definition coincides with (2) in case $X = \mathbb{R}^n$ and $K = \{x \in \mathbb{R}^n : x_i \geq 0 \text{ for } i = 1,\dots,n\}$. Now, consider two functions u,v which are continuous in $[0,a)$, differentiable in $(0,a)$ and such that

$$u(0) \leq v(0) \quad , \quad u' - f(t,u) \leq v' - f(t,v) \quad \text{in} \quad (0,a) \ .$$

Then $w(t) = v(t) - u(t)$ satisfies $w(0) \in K$ and $w' = g(t,w)$ in $(0,a)$, where

$$\text{(3)} \qquad \begin{cases} g(t,x) = f(t,u(t)+x) - f(t,u(t)) + d(t) \quad , \\ d(t) = v' - f(t,v) - (u'-f(t,u)) \quad . \end{cases}$$

Hence, if K is forward invariant with respect to g then $u(t) \leq v(t)$ in $[0,a)$. Suppose now that f is quasimonotone. Then $x^*(g(t,x)) \geq 0$ provided $x \in \partial K$, $x^* \in K^*$ and $x^*(x) = 0$, i.e.

$$\lim_{\lambda \to 0+} \lambda^{-1} \rho(x+\lambda g(t,x),K) = 0 \qquad \text{for} \quad x \in \partial K \ ,$$

by Example 4.1 . Therefore, the theorems in sect. 2 imply the following

Theorem 5.3. Let X be a real NLS , $K \subset X$ a wedge, $f: (0,a)\times X \to X$ quasimonotone with respect to K . Let u(t) and v(t) be continuous in $[0,a)$, differentiable in $(0,a)$ and such that

$$u(0) \leq v(0) \quad \text{and} \quad u'-f(t,u) \leq v'-f(t,v) \quad \text{in} \quad (0,a) \ .$$

Then $u(t) \leq v(t)$ in $[0,a)$, provided one of the following conditions is satisfied.

(i) K is a distance set and f satisfies (i) of Theorem 5.1 (with $D = K$, $\Omega = X$) .

(ii) $\overset{o}{K} \neq \emptyset$ and f is locally Lipschitz.

(iii) X is complete and f satisfies (ii) of Theorem 5.2 (with $D = K$, $\Omega = X$).

A simple result for continuously differentiable functions u,v is

Theorem 5.4. Let X be a Banach space ; $K \subset X$ a wedge ; the function $f: [0,a] \times X \to X$ be continuous and quasimonotone with respect to K ; $(f(t,x) - f(t,y), x-y)_- \leq L|x-y|^2$ for $t \in [0,a]$, and $x,y \in X$; $u,v: [0,a] \to X$ continuously differentiable and such that

$$u(0) \leq v(0) \quad \text{and} \quad u'-f(t,u) \leq v'-f(t,v) \quad \text{in} \quad [0,a] .$$

Then $u(t) \leq v(t)$ in $[0,a]$.

Proof. Consider $w(t) = v(t) - u(t)$ and the map g defined by (3). Obviously g is continuous and $(g(t,x) - g(t,y), x-y)_- \leq L|x-y|^2$. Therefore, w is the unique solution of $x' = g(t,x)$, $x(0) = w(0)$. Since g satisfies the boundary condition, w is the unique solution in K, by Theorem 4.1.

q.e.d.

4. Maximal and minimal solutions

As in finite dimensions we may combine theorems on differential inequalities and existence theorems for the initial value problem to obtain extremal solutions of this problem. At first, we prove

Lemma 5.1. Let X be a real NLS, $K \subset X$ a wedge with $\overset{o}{K} \neq \emptyset$. $D \subset X$ and $f: (0,a] \times D \to X$ quasimonotone. Let u and v be continuous in $[0,a]$, differentiable in $(0,a]$ and such that

$$u(0) < v(0) \quad \text{and} \quad u'-f(t,u) < v'-f(t,v) \quad \text{in} \quad (0,a] .$$

Then $u(t) < v(t)$ in $[0,a]$.

Proof. Let $w(t) = v(t) - u(t)$. Suppose on the contrary that $w(t) \in \overset{o}{K}$ for $0 \leq t < t_o$ and $w(t_o) \in \partial K$, for some $t_o \in (0,a]$. By Proposition 5.3 there exists $x^* \in \overset{o}{K}{}^*$ such that $x^*(w(t_o)) = 0$. Since f is quasimonotone, this implies

$$\langle f(t_o,u(t_o)) - f(t_o,v(t_o)), x^* \rangle \leq 0 .$$

By assumption we have

$$x^*(f(t,u(t)) - f(t,v(t))) > x^*(u'(t) - v'(t)) \quad \text{in} \quad (0,a] \ .$$

Hence, $x^*(u'(t_o)) < x^*(v'(t_o))$. On the other hand $x^*(u'(t_o)) \geq x^*(v'(t_o))$, since $x^*(u(t)) < x^*(v(t))$ in $[0,t_o)$. Therefore, $w(t) \in \overset{\circ}{K}$ in $[0,a]$.

q.e.d.

Theorem 5.5. Let X be a real Banach space, $K \subsetneq X$ a wedge with $\overset{\circ}{K} \neq \emptyset$, $f: [0,a] \times \overline{K}_r(x_o) \to X$ continuous and quasimonotone with respect to K , $|f(t,x)| \leq c$ and

(4) $$\alpha(f([0,a] \times B)) \leq \omega(\alpha(B)) \quad \text{for} \quad B \subset \overline{K}_r(x_o)$$

with ω from Theorem 2.2 . Let $b < \min\{a, r/c\}$. Then the initial value problem

(5) $$x' = f(t,x) \quad , \quad x(0) = x_o$$

has a maximal and a minimal (with respect to $\leq$) solution in $[0,b]$.

Proof. Let $e \in \overset{\circ}{K}$ and consider the initial value problems

(6) $$x' = f(t,x) + \frac{1}{n} e \quad , \quad x(0) = x_o + \frac{1}{n} e \ .$$

Obviously, $f_n = f + \frac{1}{n} e$ satisfies (4) . Let n be so large that

$$\frac{1}{n}|e| + (c + \frac{1}{n}|e|)b < r \ .$$

Then, by Theorem 2.2 , (6) has a solution x_n on $[0,b]$. As in the proof of Theorem 2.2 we see that (x_n) has a subsequence converging to a solution v of (5) as $n \to \infty$. On the other hand, if x is any solution of (5) then $x(t) < x_n(t)$ in $[0,b]$, by Lemma 5.1 . Therefore, $x(t) \leq v(t)$ in $[0,b]$. The existence of a minimal solution is shown similarly, by consideration of $x' = f(t,x) - \frac{1}{n} e$, $x(0) = x_o - \frac{1}{n} e$.

q.e.d.

Clearly, there is only one maximal and only one minimal solution in case K is a cone, i.e. a wedge satisfying $K \cap (-K) = \{0\}$.
Now, suppose that the hypotheses of Theorem 5.5 are satisfied and let u be continuous in $[0,b]$, differentiable in $(0,b]$ and such that

$$u(0) \leq x_o \quad \text{and} \quad u' \leq f(t,u) \quad \text{in} \quad (0,b] \ .$$

Then $u(t) \leq v(t)$ in $[0,b]$, where v is the maximal solution of (5) .

5. Remarks

(i) For $X = \mathbb{R}^n$ special cases of Theorem 5.1 have been proved by Bony [15] and Redheffer [148] who also has Proposition 5.1 for this case. Let us note that this proposition is very simple in any inner product space, since there

$$2(z,\nu) \leq \lambda|z|^2 + \lambda^{-1}\rho(x+\lambda z,D)[\rho(x+\lambda z,D) + 2|\lambda z-\nu|] \quad \text{for } \lambda > 0 .$$

By means of Lyapunov functions the results of Redheffer have been generalized in Ladde/Lakshmikantham [95]. Still in case $X = \mathbb{R}^n$, Crandall [38] has shown that $(f(x),\nu) \leq 0$ for all $x \in \partial D$ and $\nu \in N(x)$ implies $\lim_{\lambda\to 0+} \lambda^{-1}\rho(x+\lambda f(x),D) = 0$ uniformly on compact sets, provided f is continuous. Theorem 5.1 with $X = \mathbb{R}^n$ has been used by Bony [15] and Redheffer [150] to establish strong maximum principles for degenerate elliptic-parabolic equations. The following corollary to Theorem 5.1 has been proved by Brezis [18] and was useful for Nirenberg/Treves [131, Appendix].

Proposition. Let $\Omega \subset \mathbb{R}^n \times \mathbb{R}^1$ be open, $f: \Omega \to \mathbb{R}^n$ locally Lipschitz and $\phi: \Omega \to \mathbb{R}$ a C^2 function such that

(α) If $\phi(x,t_o) < 0$ then $\phi(x,t) \leq 0$ for $t \geq t_0$ with $(x,t) \in \Omega$.

(β) $f(x,t)\cdot\phi_x(x,t) \leq 0$ whenever $\phi(x,t) = 0$ (• denotes the inner product) .

(γ) $f(x,t) = 0$ whenever $\phi(x,t) = \phi_x(x,t) = \phi_t(x,t) = 0$.

Let $(x_0,t_0) \in \Omega$, $x' = f(x,t)$ in $[t_0,t_1)$ and $x(t_0) = x_0$. Then $\phi(x_0,t_0) < 0$ implies $\phi(x(t),t) \leq 0$ in $[t_0,t_1)$.

Proof. Let us write $y = (x,t) \in \Omega \subset \mathbb{R}^{n+1}$, $y(t) = (x(t),t)$ for $t \in [t_0,t_1)$. Then $y'(t) = \tilde{f}(y(t))$ where $\tilde{f}(y) = (f(y),1) \in \mathbb{R}^{n+1}$. Let D be the closure in Ω of $\{(x,t) \in \Omega$: there exists $s \leq t$ such that $\phi(x,s) < 0\}$. Obviously, $\phi(y) \leq 0$ for each $y \in D$. Let $y \in \Omega \cap \partial D$ and $\nu \in N(y)$. Then $\phi(y)=0$ and $\nu_{n+1} \leq 0$ since $(x,t) \in D$ implies $(x,t+h) \in D$ for small enough $h > 0$. Suppose first that grad $\phi(y) \neq 0$. Then $\nu = \lambda \operatorname{grad} \phi(y)$ for some $\lambda > 0$. Hence, by (β)

$$\tilde{f}(y)\cdot\nu = \lambda f(y)\cdot\phi_x(y) + \lambda\phi_t(y) \leq 0 \quad .$$

If, however, grad $\phi(y) = 0$ then $\tilde{f}(y) = (0,1)$ by (γ) . Hence,

$$\tilde{f}(y)\cdot\nu = \nu_{n+1} \leq 0 \quad .$$

q.e.d.

Some applications to matrix differential equations in particular to Riccati equations are indicated by Redheffer [149] .

(ii) Proposition 5.2 is contained in a result of Mazur [121,sect.7] . Similarly, one can show

$$(y,x)_- = -\lim_{\lambda\to 0+} \frac{|x-\lambda y|-|x|}{\lambda} = \lim_{\lambda\to 0-} \frac{|x+\lambda y|-|x|}{\lambda} \quad \text{for all } x,y\in X ,$$

in particular, $(x'(t),x(t))_- = \phi(t)D^-\phi(t)$ for $\phi(t) = |x(t)|$; cp. with Lemma 3.2 (vi) . By means of this formula, the proof of Proposition 5.1 may be simplified. In fact, since we have shown $|\nu| \leq |\nu-\lambda z|$ for every $\lambda \geq 0$, the formula implies $(z,\nu)_- \leq 0$.
Theorem 5.1 is taken from Volkmann [183] . It is a unification of results proved before by Redheffer/Walter [151] . In [152] , Redheffer/Walter have considered estimates for $\phi(x(t)-p)$ instead of $|x(t)-p|$, where ϕ is an upper semicontinuous convex functional and the conditions (i) , (ii) of Theorem 5.1 are also formulated in terms of ϕ . Theorem 5.2 (i) and Example 5.2 are taken from Volkmann [182] . It is easy to prove (i) since a simple calculation shows that $y_\varepsilon(t) = e^{-\varepsilon(t+\varepsilon)}x(t)\in \mathring{D}$ for all $t\in[0,a)$ and all $\varepsilon > 0$.
However, part (ii) which is taken from Volkmann [184] is much more difficult. Volkmann's proof depends essentially on the following fact which is a consequence of Lemma 1 in Bishop/Phelps [13] : Suppose $0\notin \overline{K}_r(x_0)$ and let K be the cone generated by $\overline{K}_r(x_0)$, i.e.

$$K = \{\lambda x : \lambda \geq 0 , x\in \overline{K}_r(x_0)\} .$$

Let $M \neq \emptyset$ be closed and bounded. Then $(z+K)\cap M = \{z\}$ for some $z\in M$. This result is used to prove

$$\overline{\lim_{\lambda\to 0+}}\ \lambda^{-1}\rho(x+\lambda f(t,x),D_\alpha) \leq \omega(t,\alpha) \quad \text{for} \quad x\in \partial D_\alpha \text{ and } t\in(0,a) ,$$

where $D_\alpha = \{x\in X : \rho(x,D) \leq \alpha\}$. Hereafter it is easy to prove (ii) . These theorems may also be proved for complex NLS , with the appropriate definition of $(\cdot,\cdot)_+$ and the wedges K , K^* of the underlying real NLS .

(iii) Definition 5.2 is due to Volkmann [181] who also has several examples showing that quasimonotonicity concepts used in earlier papers are either more restrictive, e.g. Bittner [14] and Walter [188] , or special cases. Particular cases of Theorem 5.3 and the relation between the quasimonotonicity of f and the boundary condition for g defined by (3) may already be found in Redheffer/Walter [151] . Lemma 5.1 is also proved in Volkmann [181] . Some estimation theorems related to this

lemma are given in Redheffer [147] , who has also shown that $\leq$ and $<$ may be defined by means of sublinear functionals, i.e. functionals $\phi: X \to \mathbb{R}$ such that $\phi(x+y) \leq \phi(x) + \phi(y)$ and $\phi(\lambda x) = \lambda\phi(x)$ for all $\lambda \geq 0$. To be more precise, he has shown that, given a wedge K and $\leq$ with respect to K , there exist continuous sublinear functionals ϕ such that $x \leq 0$ iff $\phi(x) \leq 0$ and $x < 0$ iff $\phi(x) < 0$. If, on the other hand, ϕ is continuous and sublinear then $K_\phi = \{x \in X : \phi(-x) \leq 0\}$ is obviously a wedge and for $\leq$ with respect to K_ϕ one has $x \leq 0$ iff $\phi(x) \leq 0$ and either $\phi(-x) \geq 0$ for all $x \in X$ or $x < 0$ iff $\phi(x) < 0$.
Maximal and minimal solutions and related comparison theorems are given in Lakshmikantham/Mitchell/Mitchell [99] .

(iv) Differential inequalities under conditions on f sufficient for existence of solutions have been considered by Martin [119] . Given a Banach space X , a closed convex set $K \subset X$ and $f,g: [0,a] \to X$ continuous, he defines $f \sim_K g$ to be equivalent with

$$(7) \quad \lim_{\lambda \to 0+} \lambda^{-1}\rho(x-y+\lambda(f(t,x)-g(t,y)),K) = 0 \quad \text{for } t \in [0,a] \text{ and } x,y \in X \text{ with } x-y \in K ,$$

and he proves the following
<u>Proposition [119, Theorem 3]</u> . Let $K \subset X$ be a wedge ; $f \sim_K f$, $f \sim_K g$ and $g \sim_K g$;

$$(f(t,x)-f(t,y),x-y)_+ \leq L|x-y|^2 \quad , \quad (g(t,x)-g(t,y),x-y)_+ \leq L|x-y|^2$$

for $t \in [0,a]$ and $x,y \in X$. Let $u,v: [0,a] \to X$ be continuously differentiable with

$$(8) \quad u(0) \leq v(0) \quad \text{and} \quad u'-g(t,u) \leq v'-f(t,v) \quad \text{in} \quad [0,a] .$$

Then $u(t) \leq v(t)$ in $[0,a]$.
Now, since K is a wedge, we notice first that $f \sim_K f$ is equivalent to the quasimonotonicity of f with respect to K , by Definition 5.2 and Example 4.1 . Next, we observe that $f \sim_K g$ is equivalent to

$$g(t,x) \leq f(t,x) \quad \text{for all } t \in [0,a] \text{ and } x \in X .$$

To see this, let x = y in (7) and notice that $x \geq 0$ iff $x^*(x) \geq 0$ for every $x^* \in K^*$. Therefore, (8) implies

$$u'-f(t,u) \leq u'-f(t,u)+f(t,u)-g(t,u) \leq v'-f(t,v) \quad \text{in } [0,a] .$$

Thus, Theorem 5.4 shows that neither $g \sim_K g$ nor

$$(g(t,x)-g(t,y),x-y)_+ \leq L|x-y|^2$$

is needed in this Proposition.

(v) The comparison theorems have been applied to the study of Cauchy's problem for parabolic equations in unbounded domains by means of semi-discretization methods ; see Walter [187, § 35] , the references given there and Lemmert [103] for the boundary layer equations. In these applications X is a weighted l^∞ space, i.e.

$$\{x \in \mathbb{R}^{\mathbb{N}} : |x| = \sup_i |x_i| \alpha_i < \infty\}$$

for some sequence (α_i) with $\alpha_i > 0$. Notice also that the matrix A corresponding to system (14) in § 4 defines a quasimonotone operator with respect to the standard cone $K = \{x \in l^1 : \inf_i x_i \geq 0\}$.

§ 6 Countable systems of ordinary differential equations

In this chapter we are concerned with the initial value problem

(1) $$x_i' = f_i(t,x_1,x_2,\dots)\ ,\ x_i(0) = c_i \quad \text{for} \quad i \in \mathbb{N}\ ,$$

where the functions $f_i\colon [0,a]\times D \to \mathbb{R}$, for some subset D of $\mathbb{R}^{\mathbb{N}}$, and the sequence $c = (c_i) \in \mathbb{R}^{\mathbb{N}}$ are given. A function $x\colon [0,b] \to D$, for some $b \in (0,a]$, is said to be a (local) solution of (1) if $x_i(0) = c_i$, $x_i \in C^1([0,b])$ and $x_i' = f_i(t,x)$ in $[0,b]$, for each $i \in \mathbb{N}$.
One way to look at problem (1) is to fix some Banach space X of real sequences and to consider (1) as $x' = f(t,x)$, $x(0) = c$ for C^1 functions with values in X . It is this point of view that we have taken in the preceding chapters. Clearly, every solution considered there is also a solution of the type mentioned above. However, in the main of the present chapter we do not worry about the Banach spaces but for conditions on the f_i and c_i such that (1) has a solution of the above type at least. Hereafter, we may ask whether such a general solution has values in some particular sequence space, since this is wanted in most applications. This approach has several advantages against the first one. Mathematically, it is the natural transition from finite to infinite systems. Secondly, problem (1) becomes less discouraging since we possibly find no solution in a particular space X but in some larger space ("generalized solutions") ; cp. Example 2.1 . We may even find solutions with values in X which cannot be obtained by means of the previous existence theorems, since the assumptions on the f_i may not ensure that the map $f = (f_1,f_2,\dots)$ satisfies the relevant conditions. Finally, let us notice that in many normed sequence spaces the existence of the (Fréchet) derivative $x'(t_0)$ implies that the components x_i are uniformly differentiable at t_0 , i.e. $x_i(t_0+h) = x_i(t_0) + hx_i'(t_0)+o(h)$ and $o(h)$ does not depend on $i \in \mathbb{N}$, a property which is not necessary for some applications.

1. Lower diagonal systems

Let us say that the system (1) is <u>lower diagonal</u> if f_n depends on the first n components only, i.e. $f_n(t,x_1,x_2,\dots) = f_n(t,x_1,\dots,x_n)$ for every $n \in \mathbb{N}$. In particular, the linear system $x' = A(t)x$ is lower diagonal iff $a_{ij}(t) \equiv 0$ for $j > i$. For example, a pure"birth"process is of this type, since $prob(S(t+h) = i \mid S(t) = j) = o(h)$ for $j > i$. Lower diagonal systems are the only ones where most things known for finite systems hold true.

Let us assume that the functions $f_n : J \times \mathbb{R}^n \to \mathbb{R}$ are continuous, where $J = [0,a]$. Then, the initial value problems

$$(1_n) \qquad x_i' = f_i(t,x_1,\dots,x_i) \quad , \quad x_i(0) = c_i \text{ for } i \leq n$$

have local solutions. Let x^n be a solution of (1_n) on $[0,b_n] \subset J$. Then we may solve

$$y' = f_{n+1}(t,x_1^n,\dots,x_n^n,y) \quad , \quad y(0) = c_{n+1}$$

on some interval $[0,\alpha]$. Therefore, $(x_1^n,\dots,x_n^n,y)$ is a solution of (1_{n+1}) on $[0,\alpha] \cap [0,b_n]$. In this way we obtain a solution of (1) provided all solutions of (1_n) exist on some interval $[0,b] \subset J$, for every $n \in \mathbb{N}$. But in general this is only a question of suitable growth conditions for the f_n and c_n ; see § 3.4 . For example, we have

Theorem 6.1. Let the system (1) be lower diagonal, $f_n : J \times \mathbb{R}^n \to \mathbb{R}$ continuous, $f^n = (f_1,\dots,f_n)$ and $|\cdot|_n$ be any norm on $\mathbb{R}^n$. Let

$$|f^n(t,x)|_n \leq \phi_n(t,|x|_n).$$

and suppose that there exists some $\alpha > 0$ such that the maximal solution ρ_n of

$$\rho' = \phi_n(t,\rho) \quad , \quad \rho(0) = |(c_1,\dots,c_n)|_n$$

exists on $[0,\alpha]$, for every $n \in \mathbb{N}$. Then (1) has a solution on $[0,\alpha] \cap J$.

In particular, the linear IVP

$$x_i' = \sum_{j \leq i} a_{ij}(t)x_j \quad , \quad x_i(0) = c_i$$

is always solvable on J provided the coefficients are continuous in J . A simple example where Theorem 1 does not apply is given by

$$f_n(t,x) = \alpha_n x_n^2 \qquad \text{and} \qquad \alpha_n c_n \to \infty \quad ,$$

since we obtain

$$x_n(t) = c_n(1-\alpha_n c_n t)^{-1} \qquad \text{in} \qquad [0,\alpha_n^{-1}c_n^{-1})$$

only, for sufficiently large n .

Concerning uniqueness it is obvious that (1) has at most one solution if every (1_n) has at most one. On the other hand it may happen that (1) has a unique solution in J but every (1_n) has several solutions in J .

Example 6.1. Let $J = [0,1]$; $f_1(x) = 2|x_1|^{1/2}$, $f_n(x) = \alpha_n x_1 x_n^2$ for $n \geq 2$; $c_1 = 0$ and $c_n > 0$ for $n \geq 2$; $\alpha_n > 0$, $(\alpha_n c_n)$ monotone and $\alpha_n c_n \to \infty$. Let (β_n) be a sequence in (0,1) such that $\alpha_n c_n(1-\beta_n)^3 < 3$. Then (1_n) has the solutions $x(t) \equiv (c_1, c_2, \ldots, c_n)$ and $y(t)$, where

$$y_1(t) = 0 \quad \text{for} \quad t \in [0,\beta_n] \quad , \quad y_1(t) = (t-\beta_n)^2 \quad \text{for} \quad t \in [\beta_n, 1]$$

$$y_i(t) = \begin{cases} c_i & \text{for } t \in [0,\beta_n] \\ 3c_i[3 - \alpha_i c_i(t-\beta_n)^3]^{-1} & \text{for } t \in [\beta_n, 1] \end{cases} \qquad (i = 2,\ldots,n).$$

Obviously, (1) has the unique solution $x(t) \equiv c$ in $[0,1]$.

Concerning comparison theorems it is obvious that they can be established in case every f^n is quasimonotone, i.e. every f_i is monotone increasing in the variables $x_1, \ldots, x_{i-1}$.
By means of Theorem 6.1 we may find solutions with values in a particular sequence space X which cannot be obtained by the previous existence theorems. Consider for example the following simple

Proposition 6.1. Let $|\cdot|_\infty$ be the usual norm of l^∞ , $J = [0,a]$, $f_n : J \times \mathbb{R}^n \to \mathbb{R}$ continuous and $|f_n(t,x)| \leq M(1+|x|_\infty)$ for every $n \in \mathbb{N}$. Let $c \in l^\infty$. Then the lower diagonal problem (1) has a solution on J with values in l^∞ , and in fact every solution has values in l^∞ .

Proof. By Theorem 6.1 , problem (1) has a solution, and if x is any solution of (1) then

$$|(x_1(t),\ldots,x_n(t))|_\infty \leq e^{Mt}(|c|_\infty + 1) - 1 \qquad \text{for every } n \in \mathbb{N} \quad .$$

q.e.d.

Example 6.2. Consider the lower diagonal linear system of a "birth" process

$$x_i' = -a_{ii}x_i + \sum_{j<i} a_{ij}x_j \quad , \quad x_i(0) = c_i \qquad \text{for } i \in \mathbb{N} \quad .$$

By Theorem 6.1 it has a unique solution x on $[0,\infty)$. Since $a_{ij} \geq 0$, the right hand sides $f_i(t,x)$ are increasing in the x_j with $j < i$, and since $c_i \geq 0$ for every i , we have $x_i(t) \geq 0$ in $[0,\infty)$ for every $i \in \mathbb{N}$. Keeping in mind that $\sum_{i>j} a_{ij} = a_{jj}$ we see that

$$\left(\sum_{i\leq n} x_i(t)\right)' \leq - \sum_{i\leq n} a_{ii}x_i(t) + \sum_{j<n} a_{jj}x_j(t) = -a_{nn}x_n(t) \leq 0.$$

Since $c \in l^1$ and $|c|_1 = 1$, this implies $x(t) \in l^1$ and $|x(t)|_1 \leq 1$ in $[0,\infty)$.

2. Row-finite systems

Let us say that system (1) is row-finite if to each $n \in \mathbb{N}$ there exists an $\alpha(n) \in \mathbb{N}$ such that

$$f_n(t,x) = f_n(t,x_1,\ldots,x_{\alpha(n)}) .$$

In particular the linear system is row-finite iff $a_{ij}(t) \equiv 0$ for $j > \alpha(i)$.
For example, systems which are obtained by semidiscretization of partial differential equations are of this type as well as systems coming from branching processes with limited"mortality". It is obvious that the results of the first section carry over to row-finite systems if $\alpha(n)>n$ only for finitely many n . However, the interesting systems are those with $\alpha(n) > n$ for every n . In this case everything is much more complicated.
Let us start with a general result for linear systems with constant coefficients. Here we need

Lemma 6.1. Let $(c_n)_{n\geq 0} \subset \mathbb{R}$. Then there exists a C^∞ function $u: \mathbb{R}^1 \to \mathbb{R}^1$ such that $u^{(n)}(0) = c_n$ for every $n \geq 0$.

Proof. Let $\psi \in C^\infty(\mathbb{R}^1)$ be such that $\psi(t) = 1$ for $|t| \leq 1/2$ and $\psi(t) = 0$ for $|t| \geq 1$. Let

$$\psi_n(t) = \frac{c_n}{n!} t^n \psi(t) ,$$

$$M_n = \max\{|\psi_n^{(k)}(t)| : t \in \mathbb{R}^1 , k < n\} \quad \text{for} \quad n \geq 1 ,$$

and choose $(\lambda_n) \subset \mathbb{R}^1$ such that $\lambda_n > 1$ for every n and $\sum_{n\geq 1} M_n\lambda_n^{-1} < \infty$.
Let us define

$$u(t) = \sum_{n \geq 0} \frac{c_n}{n!} t^n \psi(\lambda_n t) \quad [= \sum_{n \geq 0} \lambda_n^{-n} \psi_n(\lambda_n t)] \qquad \text{for} \quad t \in \mathbb{R}^1 \ .$$

We have $u \in C^\infty(\mathbb{R}^1)$ since

$$\sum_{n \geq k+1} \lambda_n^{-n} | [\psi_n(\lambda_n t)]^{(k)} | \leq \sum_{n \geq k+1} M_n \lambda_n^{-1} \qquad ,$$

and we have

$$u^{(k)}(0) = \sum_{n \geq 0} \frac{c_n}{n!} [t^n \psi(\lambda_n t)]^{(k)} |_{t=0} = c_k \qquad \text{for} \quad k \geq 0 \ .$$

q.e.d.

Theorem 6.2. Let A be row-finite. Then the linear problem

(2) $\qquad x' = Ax + b \quad , \quad x(0) = c$

has a solution in $[0,\infty)$. To be more precise, let $1 \leq n_1 < n_2 < \dots$ be the $n \in \mathbb{N}$ such that $\alpha(n_k) > n_k$, and let $\beta(k) = \max\{\alpha(n_i) : i \leq k\}$. Suppose $n_k \to \infty$ and suppose that there exists some $m \in \mathbb{N}$ such that $\beta(\beta(k)) > \beta(k)$ for every $k \geq m$. Then (2) has infinitely many solutions. In any other case, (2) has a unique solution.

Proof. It is obvious that we have a unique solution if there are only finitely many n_k . Suppose now $n_k \to \infty$ and $\beta(\beta(k)) \leq \beta(k)$ for $k = k_1, k_2, \dots$ with $k_p \to \infty$. Then the blocks

$$x_i' = (Ax)_i + b_i \quad , \quad x_i(0) = c_i \quad \text{for} \quad i \leq \beta(k_p)$$

are uniquely solvable, and therefore (1) has a unique solution. If, however, $n_k \to \infty$ and $\beta(\beta(k)) > \beta(k)$ for $k \geq m$ then at least one x_j can be choosen arbitrarily as a C^∞ function with some or all derivatives prescribed at $t = 0$. Such functions exist by Lemma 6.1 but they are not uniquely determined since, for example, addition of $\phi(t) = \exp(-t^{-2})$ does not change the derivatives at $t = 0$. The other components x_i are either uniquely determined by the x_k with $k < i$ or they may be choosen arbitrarily like x_j .

q.e.d.

Example 6.3. Let $b = 0$, $a_{i,i+1} = 1$ for $i \geq 1$ and $a_{ij} = 0$ otherwise, i.e. $x_i' = x_{i+1}$ for $i \geq 1$. The general solution of (2) is given by an arbitrary C^∞ function x_1 such that $x_1^{(n-1)}(0) = c_n$ for $n \geq 1$, and $x_n = x_1^{(n-1)}$ for $n \geq 2$.

To establish an existence theorem for nonlinear systems we need the fixed point theorem of Tychonov which is proved e.g. in [46] , [59] .

Lemma 6.2. Let X be a locally convex space, $K \subset X$ nonempty compact and convex, $T: K \to K$ continuous. Then T has a fixed point.

Theorem 6.3. Let $J = [0,a]$, $f_n: J\times\mathbb{R}^{\alpha(n)} \to \mathbb{R}$ continuous, $(c_n) \subset \mathbb{R}$,

$$|f_n(t,x)| \leq \phi_n(t,|x_1|,\dots,|x_{\alpha(n)}|) \quad \text{in } J\times\mathbb{R}^{\alpha(n)} \quad ,$$

where $\phi_n(t,\rho)$ is monotone increasing in ρ and such that the row-finite system

$$(3) \qquad \rho_n' = \phi_n(t,\rho) \quad , \quad \rho_n(0) = |c_n| \quad \text{for } n \geq 1$$

has a nonnegative solution on J . Then the row-finite problem (1) has a solution on J .

Proof. We are going to prove that the system

$$(4) \qquad u_n(t) = c_n + \int_0^t f_n(s,u(s))ds \quad \text{for } n \geq 1$$

has a continuous solution u . Then $u_n \in C^1(J)$ and we are done. Let $X = \{u : u_i \in C(J) \text{ for each } i \geq 1\}$, $|u_i|_0 = \max_J |u_i(t)|$ and

$$d(u,v) = \sum_{i\geq 1} 2^{-i} \frac{|u_i-v_i|_0}{1+|u_i-v_i|_0} \quad \text{for } u,v \in X \quad .$$

Then (X,d) is a metric locally convex space. Let ρ be the nonnegative solution of (3) and consider

$$K = \{u \in X : |u_i(t)| \leq \rho_i(t) \text{ in } J \ , \ |u_i(t)-u_i(\bar{t})| \leq |\rho_i(t)-\rho_i(\bar{t})| \text{ for } t,\bar{t} \in J \text{ and } i \geq 1\} \quad .$$

Let us define $T: K \to X$ by

$$(Tu)_i(t) = c_i + \int_0^t f_i(s,u(s))ds \quad \text{for } i \geq 1 \quad .$$

It is easy to see that $T(K) \subset K$. Since $d(u^n,u) \to 0$ iff $|u_i^n - u_i|_0 \to 0$ for every $i \geq 1$, T is continuous since the f_n are continuous. Finally, it is easy to verify that K is nonempty convex and compact. By Lemma 6.2, T has a fixed point, and this fixed point is a solution of (4) .

q.e.d.

In order to make Theorem 6.3 applicable we have to look for conditions sufficient for the existence of nonnegative solutions to the row-finite problem

(3) $$\rho_n' = \phi_n(t,\rho_1,\ldots,\rho_{\alpha(n)}) \quad , \quad \rho_n(0) = c_n > 0 \quad \text{for } n \geq 1 \; ,$$

where ϕ_n is continuous and monotone increasing in ρ , and $\phi_n(t,0) \geq 0$. In general, this is a difficult question. Clearly, the simplest condition is $\phi_n(t,\rho) \equiv M_n \geq 0$ for every $n \geq 1$. Even for linear functions ϕ_n there need not exist a nonnegative solution.

Example 6.4. Let $\phi_n(t,\rho) = \rho_{n+1}$ and $\overline{\lim_{n\to\infty}}\,(c_n/n!)^{1/n} = \infty$. By Theorem 6.2, (3) has infinitely many solutions on $[0,\infty)$, but none of them is nonnegative on any interval $[0,\alpha)$, since $\rho_n(t) \geq 0$ in $[0,\alpha)$ for every $n \geq 1$ implies

$$\rho_1(t) \;\geq\; \sum_{k\leq n} \frac{c_{k+1}}{k!}\, t^k \;\to\; \infty \qquad \text{as } n \to \infty \quad (\text{for } t > 0) \; .$$

We are not able to prove Theorem 6.2 for systems with nonconstant coefficients ; for continuous a_{ij} and b_i we have for instance

$$\phi_n(t,\rho) \;=\; M_n(1 + \sum_{i\leq\alpha(n)} \rho_i)$$

with $M_n = \max\{|a_{ni}(t)|,|b_n(t)| \;:\; i \leq \alpha(n) \;,\; t \in J\}$. But Example 6.4 shows that the corresponding problem (3) need not have a nonnegative solution and therefore Theorem 6.3 does not apply. Perhaps a suitable generalization of Lemma 6.1 will give existence. Particular linear systems will be considered in sect. 4 .
The following simple nonlinear example shows that the usual comparison theorems are no longer valid for row-finite systems, even when $f(t,\cdot)$ maps l^∞ into l^∞ , for instance.

Example 6.5. Let $J = [0,1]$, $(t_n) \subset (0,1)$ be strictly decreasing, $\alpha_n(t) = 0$ for $t \in [0,t_{n+1}]$ and positive for $t \in (t_{n+1},1]$,

$$f_n(t,x) \;=\; -\,4|x_n|^{2/3} + \alpha_n(t)x_{n+1} \quad .$$

Obviously, f_n is monotone increasing in x_{n+1} . Consider $u(t) \equiv 0$ and $v_n(t) = (t_n-t)^3$ for $n \geq 1$. It is easy to verify that

$$u_i' - f_i(t,u) = 0 < v_i' - f_i(t,v) \text{ in } [0,1] \text{ for each } i \geq 1,$$

but we do not have $u(t) \leq v(t)$ in J .

3. General systems

The situation described for row-finite systems becomes even worse for general systems since we now have convergence problems in the rows too. For instance, we have already shown in Example 1.2 that the linear problem $x' = Ax$, $x(0) = c$ need not have a solution if at least one row of A has infinitely many nonvanishing elements.
The following theorem is the extension of Theorem 6.3 to general systems.

Theorem 6.4. Let $J = [0,a]$, $\mathbb{R}^{\mathbb{N}}$ be given the topology generated by the metric

$$d(x,y) = \sum_{i \geq 1} 2^{-i} \frac{|x_i - y_i|}{1+|x_i - y_i|} \qquad \text{for} \quad x,y \in \mathbb{R}^{\mathbb{N}} ,$$

$f_n : J \times \mathbb{R}^{\mathbb{N}} \to \mathbb{R}$ continuous, $(c_n) \subset \mathbb{R}$ and $|f_n(t,x)| \leq \phi_n(t,|x_1|,|x_2|,\ldots)$ in $J \times \mathbb{R}^{\mathbb{N}}$, where $\phi_n(t,\rho)$ is monotone increasing in ρ and such that

$$\rho_n' = \phi_n(t,\rho) \quad , \quad \rho_n(0) = |c_n| \qquad \text{for} \quad n \geq 1$$

has a nonnegative solution on J . Then problem (1) has a solution on J.

This theorem can be proved like Theorem 6.3 . Notice that the continuity of f_n with respect to the metric d is equivalent to continuity with respect to the product topology of $\mathbb{R}^{\mathbb{N}}$, and in general this is a strong condition. For linear functions

$$f_n(x) = \sum_{j \geq 1} a_{nj} x_j$$

this continuity condition is satisfied iff $a_{nj} \neq 0$ for at most finitely many j , i.e. the matrix $A = (a_{ij})$ must be row-finite. Therefore, let us state the following

Corollary 6.1. Let $J = [0,a]$; $D = \{x \in \mathbb{R}^{\mathbb{N}} : |x_i - c_i| \leq r_i \text{ for each } i \in \mathbb{N}\}$ for some $(r_i) \in \mathbb{R}^{\mathbb{N}}$ with $r_i > 0$ for each $i \geq 1$; $f_i : J \times D \to \mathbb{R}$ continuous (with respect to the metric d) , $M_i = \max_{J \times D} |f_i(t,x)|$ and $M_i \leq M r_i$ for some constant $M > 0$ and every $i \geq 1$. Then (1) has a solution in $[0,1/M] \cap J$.

This result follows from Theorem 6.4 by continuous extension of the f_i to $J\times\mathbb{R}^{\mathbb{N}}$. Notice that $M_i < \infty$ since D is compact with respect to d. For Lipschitz continuous functions f_i we have

<u>Corollary 6.2</u>. Let J and $D \subset \mathbb{R}^{\mathbb{N}}$ be as in Corollary 6.1 ; $f_i : J\times D \to \mathbb{R}$ continuous in t and

$$|f_i(t,x) - f_i(t,y)| \leq \sum_{j\geq 1} L_{ij}|x_j - y_j| \qquad \text{for } x,y \in D$$

with

$$\sum_{j\geq 1} L_{ij} r_j < \infty \qquad \text{for all } i \geq 1 \quad ;$$

$M_i = \max_{J\times D} |f_i(t,x)|$, and $M_i \leq Mr_i$ for some $M > 0$ and every $i \geq 1$. Then (1) has a solution in $[0,1/M] \cap J$. If in addition

$$\sum_{j\geq 1} L_{ij} r_j \leq \tilde{M} r_i$$

for some $\tilde{M} > 0$ and every $i \geq 1$, then (1) has exactly one solution on $[0,1/M] \cap J$.

<u>Proof</u>. Since $\sum_{j\geq 1} L_{ij} r_j < \infty$, f_i is continuous with respect to d . Hence, $M_i < \infty$. Therefore, Corollary 6.1 applies. Now, let x and y be solutions of (1) in $J_o = [0,1/M] \cap J$, and let $w_i = |x_i - y_i|/r_i$. Then $w_i(t) \leq 2$ in J_o and

$$w_i(t) \leq r_i^{-1} \sum_{j\geq 1} L_{ij} r_j \int_0^t w_j(s)ds \qquad \text{for } i \geq 1 \quad .$$

Now, suppose that

$$r_i^{-1} \sum_{j\geq 1} L_{ij} r_j \leq \tilde{M} \qquad \text{for each } i \geq 1 \quad ,$$

and let $\phi(t) = \sup_i w_i(t)$. Since $|\phi(t) - \phi(\bar{t})| \leq 2M|t-\bar{t}|$, ϕ is continuous and we have

$$\phi(t) \leq \tilde{M} \int_0^t \phi(s)ds \quad .$$

Therefore, $\phi(t) \equiv 0$ in J_o. q.e.d.

Let us consider a simple application.

<u>Example 6.6</u>. In the theory of neural nets one was led to consider the nonlinear system

$$\alpha x_i' + \alpha_i x_i = \frac{1}{1+\exp[-\gamma_i - \sum_{j\geq 1} \beta_{ij} x_j]} \equiv f_i(x) \quad , \; x_i(0)=c_i \text{ for } i \geq 1 \; ;$$

see e.g. [134] and the references given there. The function $x_i(t)$ represents the sensitivity of the ith cell and has its range in $[0,1]$. The constants $\alpha > 0$, $\alpha_i \geq 0$, $\gamma_i \geq 0$ and β_{ij} are given.
We may assume $\alpha = 1$, considering $y_i(t) = x_i(\alpha t)$ if necessary. Then we have

$$(5) \qquad \begin{cases} x_i' = -\alpha_i x_i + f_i(x) \\ x_i(0) = c_i \in [0,1] \end{cases} \qquad \text{for } i \geq 1 \quad .$$

Suppose x is a solution of (5) . Then

$$0 \leq (x_i e^{\alpha_i t})' \leq e^{\alpha_i t} \quad ,$$

and therefore

$$c_i e^{-\alpha_i t} \leq x_i(t) \leq c_i e^{-\alpha_i t} + t \quad .$$

Let $J = [0,r]$, and let us assume that

$$\sum_{j\geq 1} |\beta_{ij}| < \infty \qquad \text{for each } i \geq 1 \quad .$$

Then we may apply either Theorem 6.4 or Corollary 6.2 . Since we have a priori bounds for the solution, let us apply Theorem 6.4 . We define $D = \{x \in \mathbb{R}^{\mathbb{N}} : 0 \leq x_i \leq c_i+r \text{ for each } i \geq 1\}$, $\tilde{f}_i : \mathbb{R}^{\mathbb{N}} \to \mathbb{R}$ by $\tilde{f}_i(x)=f_i(x)$ for $x \in D$ and by the corresponding boundary values of f when $x \notin D$, i.e. for example

$$\tilde{f}_i(x) = f_i(x_1,\ldots,x_{j-1},c_j+r,x_{j+1},\ldots)$$

if $x_j > c_j+r$ but $x_k \in [0,c_k+r]$ for $k \neq j$. Clearly

$$|-\alpha_i x_i + \tilde{f}_i(x)| \leq \alpha_i |x_i| + 1$$

and the majorizing system $\rho_i' = \alpha_i \rho_i + 1$, $\rho_i(0) = c_i$ has a nonnegative solution on J . Moreover, $\tilde{f}_i$ is continuous with respect to d . Therefore (5) has a solution on J . Certainly, we have sensitivities in $0 \leq t \leq 1 - |c|_\infty$, i.e. $0 \leq x_i(t) \leq 1$ for every $i \geq 1$, provided $|c|_\infty < 1$.

4. The function exp(At)

Let X be a Banach space of real sequences and suppose that the matrix $A = (a_{ij})$ defines a bounded linear operator from X into X . Then we know that

$$e^{At} = \sum_{n \geq 0} \frac{t^n}{n!} A^n$$

exists for each $t \geq 0$, and the solution of the linear problem

$$(6) \qquad x_i' = \sum_{j \geq 1} a_{ij} x_j \quad , \quad x_i(0) = c_i \quad \text{for} \quad i \geq 1$$

is given by $x(t) = e^{At}c$ for $c \in X$ and $t \in [0,\infty)$. Moreover, $c_i \geq 0$ and $a_{ij} \geq 0$ for each i and j imply $x_i(t) \geq 0$ on $[0,\infty)$. In this section, we consider such a representation and its consequences for more general matrices A .

Definition 6.1. Let $A = (a_{ij})$ be an infinite matrix and suppose that there is some $r > 0$ such that the power series

$$\sum_{n \geq 0} \frac{t^n}{n!} (A^n)_{ij}$$

has radius of convergence $\geq r > 0$, for each $i,j \in \mathbb{N}$. Then we define the matrix e^{At} by

$$(e^{At})_{ij} = \sum_{n \geq 0} \frac{t^n}{n!} (A^n)_{ij} \quad .$$

Furthermore, we let $|A| = (|a_{ij}|)$ and $|c| = (|c_i|)$ for $c \in \mathbb{R}^{\mathbb{N}}$.

It is easy to see that e^{At} exists in $[0,\infty)$ and is lower diagonal provided A is lower diagonal. Moreover, $e^{At}c$ is the solution of the lower diagonal problem (6) , for every $c \in \mathbb{R}^{\mathbb{N}}$. The following simple example shows that e^{At} may not exist if A is only row-finite.

Example 6.7. Let $a_{i+1,i} = a_{i,i+1} = i^{1+\alpha}$ for some $\alpha > 0$, and $a_{ij} = 0$ otherwise. Let us show that e^{At} is not defined. We have

$$(A^{2p})_{ii} > a_{i,i+1}a_{i+1,i+2}\cdots a_{i+p-1,i+p}a_{i+p,i+p-1}\cdots a_{i+1,i}$$

$$= \left[\prod_{k=0}^{p-1} (i+k)\right]^{2(1+\alpha)} \quad .$$

Hence,

$$\sum_{n\geq o} \frac{t^n}{n!}(A^n)_{ii} \geq \sum_{p\geq 1} \frac{t^{2p}}{(2p)!}(A^{2p})_{ii} \geq \sum_{p\geq 1} \frac{t^{2p}}{(2p)!}\left[\frac{(p+i-1)!}{(i-1)!}\right]^{2(1+\alpha)}$$

$$= [(i-1)!]^{-2(1+\alpha)} \sum_{p\geq 1} \frac{[(p+i-1)!]^{2(1+\alpha)}}{(2p)!} t^{2p} = \infty$$

for $t > 0$.

Theorem 6.5. Let $\exp(|A|t)$ be defined in $[0,r)$, $\alpha \in (0,r)$ and

$$D_\alpha = \{x \in \mathbb{R}^{\mathbb{N}} : \sum_{j\geq 1} (e^{|A|\alpha})_{ij}|x_j| < \infty \text{ for every } i \geq 1\} .$$

Let $c \in D_\alpha$ and $b \in D_\alpha$. Then

$$x(t) = e^{At}c + \int_0^t e^{A(t-s)}b\,ds \qquad \text{for } t \in [0,\alpha]$$

is a solution of $x' = Ax+b$, $x(0) = c$. Moreover, the successive approximations,

$$x^{n+1}(t) = c + \int_0^t \{Ax^n(s)+b\}ds \qquad \text{for } n \geq 0 , \quad x^o(t) \equiv c$$

converge to $x(t)$ on $[0,\alpha]$.

Proof. Since all series are absolutely convergent in $[0,\alpha]$, it is obvious that x is a solution, and since

$$x^n(t) = \sum_{k=o}^{n} \frac{t^k}{k!} A^k c + \sum_{k=1}^{n} \frac{t^k}{k!} A^{k-1} b \qquad \text{for } n \geq 1 ,$$

we have $x^n(t) \to x(t)$ in $[0,\alpha]$.

q.e.d.

Clearly, it would have been enough to assume that A , c and b are such that x^n are defined in $[0,\alpha]$ and that they converge to x . But in concrete applications it would be hard to verify such conditions on A . Consider, however, the following simple test for the existence of $\exp(|A|t)$.

Lemma 6.3. Let $A = (a_{ij})$ and suppose there exists a matrix (b_{pq}) with $b_{pq} \geq 0$ for $p,q \geq 0$ such that

(i) $b_{po} \geq 1$ for $p \geq 0$ $\quad$ [or $b_{oq} \geq 1$ for $q \geq 0$]

(ii) $\sum_{j\geq 1} |a_{ij}|b_{jk} \leq b_{i,k+1}$ $\quad$ [or $\sum_{j\geq 1} b_{ij}|a_{jk}| \leq b_{i+1,k}$]

(iii) $\sum_{q\geq 0} \frac{b_{pq}}{q!} t^q$ has radius $r > 0$ for every $p \geq 1$

[or $\sum_{p\geq 0} \frac{b_{pq}}{p!} t^p$ has radius $r > 0$ for every $q \geq 1$] .

Then $\exp(|A|t)$ is defined on $[0,r)$. In addition,

$$\sum_{j\geq 1} (e^{|A|t})_{ij} < \infty \quad \text{for } i \geq 1 \quad \left[\text{or } \sum_{i\geq 1} (e^{|A|t})_{ij} < \infty \quad \text{for } j \geq 1\right].$$

<u>Proof</u>. Let us assume that [...] holds in (i) - (iii) . By induction on n we shall show that

$$\sum_{i\geq 1} (|A|^n)_{ij} \leq b_{nj} \quad .$$

Then, the assertion [...] follows from (iii) .
For $n = 0$ we have $\sum_{i\geq 1} \delta_{ij} = 1 \leq b_{0j}$. For $n = 1$ we have

$$\sum_{i\geq 1} |a_{ij}| \leq \sum_{i\geq 1} b_{0i}|a_{ij}| \leq b_{1j} \quad ,$$

by (i) and (ii) . Suppose the result is true for n . Then

$$\sum_{i\geq 1} (|A|^{n+1})_{ij} = \sum_{i\geq 1} \sum_{k\geq 1} (|A|^n)_{ik}|a_{kj}| \leq \sum_{k\geq 1} b_{nk}|a_{kj}| \leq b_{n+1,j} \quad .$$

q.e.d.

The following example contains some special cases.

<u>Example 6.8</u>. (a) Suppose

$$M = \sup_i \sum_{j\geq 1} |a_{ij}| < \infty \qquad \left[\text{or } \sup_j \sum_{i\geq 1} |a_{ij}| < \infty\right] \quad .$$

Then $r = \infty$ in Lemma 6.3 and we may choose $b_{pq} = (1+M)^q$.

(b) Suppose there exists some $N \geq 1$ such that $a_{ij} = 0$ for $j > N$ [or $a_{ij} = 0$ for $i > N$] . Then $r = \infty$.

(c) Suppose A is lower diagonal. Then $r = \infty$ and we may choose $b_{pq} = (1+M_p)^q$ with

$$M_p = \max_{m\leq p} \sum_{k\geq 1} |a_{mk}| \quad .$$

(d) Suppose there exist an index N and a constant $M > 0$ such that $a_{ij} = 0$ for $j > i+N$ and

$$(*) \qquad \sum_{j\geq 1} |a_{ij}| \leq M\cdot i \quad \text{for every} \quad i \geq 1$$

[or $a_{ij} = 0$ for $i > j+N$ and $\sum_{i\geq 1} |a_{ij}| \leq M\cdot j$ for every $j \geq 1$] .

Let $b_{po} = 1$ and

$$b_{pq} = M^q \prod_{k=o}^{q-1} (p+kN) \quad \text{for} \quad q \geq 1 .$$

Then (ii) is satisfied since

$$\sum_{j=1}^{i+N} |a_{ij}| b_{jk} \leq M^k \sum_{j=1}^{i+N} |a_{ij}| \prod_{m=1}^{k} (i+mN)$$

$$\leq M^k \prod_{m=1}^{k} (i+mN)Mi = b_{i,k+1} .$$

Since $\sum_{q\geq o} b_{pq} \frac{t^q}{q!}$ has radius $(NM)^{-1}$, $\exp(|A|t)$ is defined in $[0,\rho)$ with $\rho \geq (NM)^{-1}$. Example 6.7 shows that the condition (*) above is sharp.

Evidently, the solution x given by Theorem 6.5 need not be the only one ; see Example 6.3 . But if we know that $a_{ij} \geq 0$, $c_i \geq 0$ and $b_i \geq 0$ then we have $x_i(t) \geq 0$ in $[0,\alpha]$ for every $i \geq 1$, and we shall see in the next chapter that this property of x remains true if the a_{ii} are negative. Therefore, such linear systems are admissible candidates for the majorizing systems in Theorem 6.3 and Theorem 6.4 . In particular, we may obtain existence theorems for linear systems with variable coefficients.

5. Remarks

(i) Lemma 6.1 is due to E. Borel [16] . Actually, he proved that, given $\delta > 0$, there exists a real power series $u(t)$, convergent in $|\delta-t| < \delta$ such that $\lim_{t\to 0+} u^{(n)}(t) = c_n$ for every $n \geq 0$. Another proof is given in Treves [172,p.390] .
Example 6.5 is taken from Walter [189] .

(ii) Theorem 6.4 has been proved by A.N. Tychonov [173] , [174] in case $\phi_n(t,\rho) \equiv M_n$. These papers have been the starting point for a series

of investigations by Russian mathematicians, many of them published in Izvestija Akad. Nauk Kazach. SSR; see e.g. Persidskii [137], [138], [139], Bagautdinov [5], Zautykov [198] and the book of Valeev/Zautykov [199].
In addition to the assumptions of Theorem 6.4 (or the corollaries), assume that every f_i is increasing in x_j for every $j \neq i$. Then it is obvious that (1) has a maximal solution $\hat{x}$, and $y_i(0) \leq c_i$ together with $D^+y_i(t) \leq f_i(t,y(t))$ for each $i \geq 1$ implies $y_i(t) \leq \hat{x}_i(t)$ for each $i \geq 1$. In case $\phi_n(t,\rho) \equiv M_n$ this has been shown by Mlak/Olech [125] and repeated in Mlak [124].
The second part of Corollary 6.2 has been established by Hart [71] in 1917, by means of successive approximations of course. Some remarks on other early papers on countable systems will be given in the next chapter.

(iii) Example 6.6 is taken from Oguztöreli [134], who has (α_i), $(\gamma_i) \in [0,1]$ and $\sup_i \sum_{j\geq 1} |\beta_{ij}| < \infty$. The stability of the equilibrium, i.e. the solution of the system with $x_i' \equiv 0$ for every $i \geq 1$, is discussed in Leung/Mangeron/Oguztöreli/Stein [104].

(iv) Example 6.7, Lemma 6.3 and Example 6.8 are taken from Arley/Borchsenius [4], who have applications to branching processes and to the perturbation theory of quantum mechanics (Fourier method) as well as several pathological examples for general systems that do not satisfy the conditions of Lemma 6.3. Some related examples have been considered later in Hille [75], [76].

(v) Clearly, the existence of $\exp(|A|t)$ does not imply that all rows [or all columns] of this matrix are in l^1, as it is the case in Lemma 6.3. Consider the simple

Example 6.9. Let $a_{ij} = \alpha_i\beta_j$ with $\alpha_i \geq 0$, $\beta_i \geq 0$ and

$$\alpha = \sum_{i\geq 1} \alpha_i\beta_i < \infty .$$

Then $A^n_{ij} = \alpha_i\beta_j\alpha^{n-1}$ for $n \geq 1$ and therefore

$$e^{At}_{ij} = \alpha_i\beta_j\alpha^{-1}(e^{\alpha t} - 1) + \delta_{ij} .$$

If we choose, for example, $(\alpha_i) \in l^p \backslash l^1$ and $(\beta_i) \in l^q \backslash l^1$ with $p > 1$ and $p^{-1} + q^{-1} = 1$, then no row and no column of $\exp(At)$ belongs to l^1.

Now, let $A = |A|$, suppose that $\exp(At)$ exists for some $t > 0$ and let

$(a_{ij})_{j\geq 1}$ belong to the sequence space X . Then every column $(a_{ij})_{i\geq 1}$ must be in the α-dual $X^{\times}$ of X , i.e. in the space

$$\{x \in \mathbb{R}^N : \sum_{j\geq 1} |x_j| a_{ij} < \infty\} \qquad ;$$

see Koethe [88.§ 30] . In Lemma 6.3 we have $X = l^1$. It would be interesting to have a criterion like Lemma 6.3 in this more general setting.

(vi) In the theory of dissociation of polymers one was led to consider the system

$$(7) \qquad x_i' = -(i-1)a_{i-1,i}x_i + 2\sum_{j\geq i+1} a_{ij}x_j \quad , \quad x_i(0) = c_i \quad \text{for } i \geq 1 \; ,$$

where $x_i(t) \geq 0$ denotes the concentration of i-mers in a mixture of polymers of various degrees of polymerization with the same monomers, and the a_{ij} are nonnegative constants. The first term of the right hand side in (7) represents the rate of i-mers which disappear by dissociation into shorter chains while the second term comes from the dissociation of j-mers with $j > i$ into i-mers. Furthermore, the solution x of (7) has to satisfy the side condition

$$(8) \qquad \sum_{i\geq 1} i\, x_i(t) \equiv K_o \quad ,$$

where the constant K_o is the total number of monomers per unit volume. Oguztöreli [133] has proved local existence of a solution to (7) in case $c \in l^\infty$ and the strong condition

$$\sum_{i,j\geq 1} i\, a_{ij} < \infty$$

holds. Obviously, this is a consequence of Example 6.8 (a) , and in fact the solution exists in $[0,\infty)$ and is the only one with values in l^∞ . Concerning solutions of (7) satisfying (8) he assumes in addition that

$$\sum_{i\geq 1} i(i-1)a_{i-1,i} < \infty \quad \text{and} \quad (i-1)(i-2)a_{i-1,i} = 2\sum_{j=1}^{i-2} j\, a_{ji}$$

to obtain a unique solution by means of successive approximations. Clearly, the necessary condition $\sum_{i\geq 1} i\, c_i = K_o$ has also to be satisfied.

The original system proposed by Simha [163] , where $a_{ij} = \alpha > 0$ for all $i,j \geq 1$, does not meet any of these requirements ; Simha has only solved the finite dimensional case, i.e. he assumed that $x_j \equiv 0$ for

all $j \geq n+1$ for some n . In the next chapter we shall solve the countable system.
Hille [75] , [76] has considered systems of the type

$$(9) \qquad x_n' + na_n x_n = \sum_{j \geq n+1} a_j x_j \qquad \text{for} \quad n \geq 1$$

in the space $X = \{x \in \mathbb{R}^{\mathbb{N}} : |x| = \sum_{n \geq 1} a_n |x_n| < \infty\}$, where the $a_n > 0$ are given constants. He has shown that (9) has a nontrivial solution x with values in X and $x(0) = 0$ provided $\sum_{n \geq 1} 1/(n^2 a_n) < \infty$. In case $a_n = 1$ for each n , it is possible to determine the general solution of (9) explicitely in terms of an arbitrary locally integrable function.

(vii) It is useful to keep in mind that the initial value problem is sometimes equivalent to the corresponding infinite system of integral equations. For example, consider the linear problem $x' = Ax$, $x(0) = c$, where $c \in l^1$ and the complex numbers a_{ij} are such that

$$\sup_i \operatorname{Re} a_{ii} < \infty \quad \text{and} \quad \sum_{i \geq 1, i \neq j} |a_{ij}| \leq M \quad \text{for all} \quad j \geq 1 .$$

By means of semigroup theory, McClure/Wong [122] have shown that the IVP has exactly one continuous solution $x: [0,\infty) \to l^1$. In fact this result is a simple consequence of Banach's fixed point theorem: Consider any $a > 0$, $J = [0,a]$, $X = C_{l^1}(J)$ with

$$|x| = \max_J \{|x(t)|_{l^1} e^{-\alpha t}\} \qquad \text{for some} \quad \alpha > 0$$

and the equivalent system

$$x_i(t) = c_i e^{a_{ii}t} + \int_0^t e^{a_{ii}(t-s)} \sum_{j \neq i} a_{ij} x_j(s)ds = (Tx)_i(t) \text{ for } i \geq 1.$$

By splitting infinite sums into finite sums and remainders, it is easy to see that T maps X into itself, and T is a strict contraction provided α is sufficiently large.

(viii) An interesting application of existence and comparison theorems to countable systems obtained by semidiscretization of Cauchy's problem for parabolic equations is given in Voigt [180] .

§ 7 Approximate solutions

In the first chapter we have obtained approximate solutions for the initial value problem

$$(1) \qquad x' = f(t,x) \quad , \quad x(0) = c$$

by approximating the continuous function f by functions which are locally Lipschitz. In the fourth chapter we have constructed approximate solutions by means of the Euler-Cauchy polygon method. Other classical methods are that one of Peano, where the approximate solutions x_ε for $\varepsilon > 0$ are defined by

$$x_\varepsilon(t) = \begin{cases} c & \text{for } t \leq 0 \\ c + \int_0^t f(s,x_\varepsilon(s-\varepsilon))ds & \text{for } t \geq 0 \end{cases} ,$$

and the method of successive approximation. We notice that all these approximate solutions have values in the infinite dimensional Banach space X . However, for numerical purposes it is highly desirable to approximate solutions of (1) by solutions of finite dimensional systems, since there are many effective methods to compute such approximate solutions, approximately. One way to do this consists in the consideration of finite dimensional subspaces X_n of X , projections P_n of X onto X_n and the finite system

$$(2) \qquad x' = P_n f(t,x) \quad , \quad x(0) = P_n c \quad \text{for } x \in X_n \quad .$$

In particular, suppose X has a Schauder base (e_i, e_i^*) . Then we may consider the span of $\{e_1,\dots,e_n\}$ for X_n and the natural projection P_n , defined by

$$P_n x = \sum_{i \leq n} \langle x, e_i^* \rangle e_i \quad .$$

For example, if we have $X = l^1$ and the natural base of l^1 then (2) becomes

$$(1_n) \qquad x_i' = f_i(t,x_1,\dots,x_n,0,0,\dots) \quad , \quad x_i(0) = c_i \quad \text{for } i \leq n.$$

Now, let us forget the special Banach space and let us consider the general countable system (1) . Then we may still use (1_n) to obtain approximate solutions. In our days such an approach is commonly related to the name Galerkin, and therefore we shall say that the solutions of (2) or (1_n) are Galerkin approximations to the solutions of (1) , while more than sixty years ago, F. Riesz [156] spoke of the "principe des réduites" .

1. Galerkin approximations in a Banach space

Let X be a real Banach space such that there exists a sequence of finite dimensional subspaces $X_n \subset X$ and a sequence of linear projections P_n from X onto X_n with $|P_n| = 1$ for each $n \geq 1$ and $P_n x \to x$ for each $x \in X$.
The simplest example of such a space is a separable Hilbert space, where X_n and P_n are defined by means of an orthonormal base. Other examples may be found e.g. in [46, Chap.6] .
We want to show, under hypotheses similar to those in Theorem 3.2 , that the Galerkin approximations (2) converge to the solution of (1) . To this end we need

<u>Proposition 7.1</u>. Let X , X_n and P_n be as above. Then

$$(P_n x,y)_- \leq (x,y)_+ \quad \text{for} \quad x \in X \quad \text{and} \quad y \in X_n \ .$$

<u>Proof</u>. We know that $(P_n x,y)_- = y^*(P_n x) = P_n^* y^*(x)$ for some $y^* \in Fy$. Since $|P_n^*| = 1$, we have $|P_n^* y^*| \leq |y^*| = |y|$ and $P_n^* y^*(y) = y^*(y) = |y|^2$, i.e. $P_n^* F(y) \subset F(y)$. Therefore, $(P_n x,y)_- \leq (x,y)_+$.

q.e.d.

<u>Theorem 7.1</u>. Let X be a real Banach space with a projectional scheme $\{X_n,P_n\}$ such that $|P_n| = 1$ for each n and $P_n x \to x$ for each $x \in X$. Let $f\colon [0,a]\times\overline{K}_r(c) \to X$ be continuous and bounded, say $|f(t,x)| \leq M$, and

$$(3) \quad (f(t,x)-f(t,y),x-y)_+ \leq \omega(t,|x-y|)|x-y| \quad \text{for } t \in (0,a] \ ; \ x,y \in \overline{K}_r(c),$$

where ω is of class U_1 (cp. sec. 3.3) . Let $b < \min\{a,r/M\}$.
Then problem (1) has a unique solution x on $[0,b]$, problem (2) has a unique solution x_n on $[0,b]$ for sufficiently large n , and $x_n(t) \to x(t)$ as $n \to \infty$, uniformly on $[0,b]$.

Proof. By Theorem 3.2 , problem (1) has a unique solution. Since $P_n c \to c$ and $|P_n| = 1$, we may choose n_o so large that $|P_n c - c| + bM \leq r$ for $n \geq n_o$. Hence, (2) has a solution on $[0,b]$. It is uniquely determined since $P_n f(t,x)$ for $x \in X_n \cap \overline{K}_r(c)$ satisfies (3) , in consequence of Proposition 7.1 . To prove $x_n(t) \to x(t)$, let $y_n(t) = P_n x(t)$, $z_n(t) = x_n(t) - y_n(t)$ and $\phi(t) = |z_n(t)|$. Then we have

$$\begin{aligned}\phi(t)D^-\phi(t) &\leq (z_n', z_n)_- \\ &\leq (P_n f(t,x_n) - P_n f(t,y_n), z_n)_- + |f(t,y_n) - f(t,x)|\phi(t) \\ &\leq \omega(t,\phi(t))\phi(t) + |f(t,P_n x(t)) - f(t,x(t))|\phi(t)\end{aligned}$$

in $(0,b]$ and

$$|f(t,P_n x(t)) - f(t,x(t))| \to 0 \quad \text{as} \quad n \to \infty \quad ,$$

uniformly in $[0,b]$. Furthermore,

$$\frac{\phi(t)}{t} \to |P_n f(0,P_n c) - P_n f(0,c)| = \alpha_n \quad \text{as} \quad t \to 0+ \quad ,$$

and $\alpha_n \to 0$ as $n \to \infty$. Hence, we may continue as in the proof of Theorem 3.2 to obtain $z_n(t) \to 0$, and since $y_n(t) \to x(t)$, we have $x_n(t) \to x(t)$, uniformly on $[0,b]$.

q.e.d.

In order to establish a related theorem, under hypotheses similar to those in Theorem 2.2 , we need

Proposition 7.2. Let X be a real Banach space with a projectional scheme $\{X_n, P_n\}$ such that $P_n x \to x$ for each $x \in X$. Let γ be the (Hausdorff-) measure of noncompactness for X , and $B \subset X$ be bounded. Then

$$\gamma(B) \leq \gamma(\bigcup_{n \geq 1} P_n B) = \lim_{m \to \infty} \gamma(\bigcup_{n \geq m} P_n B) \leq \gamma(B) \sup_n |P_n| \quad .$$

Proof. The second equality is obvious since $P_n B$ is relatively compact. Let

$$\bigcup_{n \geq m} P_n B \subset \bigcup_{i \leq k} \overline{K}_r(x_i) \qquad \text{with} \qquad r > \gamma(\bigcup_{n \geq m} P_n B) \quad .$$

Then $B \subset \bigcup_{i \leq k} \overline{K}_r(x_i)$, since $P_n x \to x$ for each $x \in X$, and therefore $\gamma(B) \leq r$. Hence,

$$\gamma(B) \leq \lim_{m \to \infty} \gamma(\bigcup_{n \geq m} P_n B) \quad .$$

Now, let $B \subset \bigcup_{i \le k} K_r(x_i)$ with $r > \gamma(B)$. Given $\varepsilon > 0$, we find an index m such that $|P_n x_i - x_i| \le \varepsilon$ for $n \ge m$ and $i \le k$. Therefore,

$$\bigcup_{n \ge m} P_n B \subset \bigcup_{i \le k} K_{\lambda r + \varepsilon}(x_i) \qquad \text{with} \qquad \lambda = \sup_n |P_n|$$

This implies $\lim_{m \to \infty} \gamma(\bigcup_{n \ge m} P_n B) \le \gamma(B) \sup_n |P_n|$.

q.e.d.

Theorem 7.2. Let X be a Banach space with a projectional scheme $\{X_n, P_n\}$ such that $P_n x \to x$ for each $x \in X$. Let $f\colon [0,a] \times \overline{K}_r(c) \to X$ be continuous and bounded, say $|f(t,x)| \le M$, and

$$\gamma(f(J \times B)) \le \omega(\gamma(B)) \qquad \text{for} \qquad B \subset \overline{K}_r(c) \qquad ,$$

where ω is of the same type as in Theorem 2.2 . Let $b < \min\{a, r/M\}$. Then there exists an index n_0 such that (2) has a solution x_n on $[0,b]$, for $n \ge n_0$. The sequence (x_n) has a subsequence that converges to a solution of (1) , uniformly on $[0,b]$. In case (1) has a unique solution x on $[0,b]$, the whole sequence $(x_n)_{n \ge n_0}$ converges to x .

Proof. Obviously, (2) has a solution x_n which is defined on $[0,b]$ provided n is sufficiently large. As in the proof of Theorem 2.2 , let $B_k(t) = \{x_n(t) : n \ge k\}$ and $\phi_k(t) = \gamma(B_k(t))$. We have

$$D^- \phi_k(t) \le \overline{\lim_{\tau \to 0+}}\ \gamma(\bigcup_{J_\tau} B_k'(s)) \qquad \text{for} \quad t \in (0,b] \quad \text{and} \quad J_\tau = [t-\tau, t] \quad ,$$

and

$$\gamma(\bigcup_{J_\tau} B_k'(s)) \le \gamma(\bigcup_{n \ge k} P_n f(J \times \bigcup_{J_\tau} B_k(s))) \le \lambda \gamma(f(J \times \bigcup_{J_\tau} B_k(s))$$

$$\le \lambda \omega(\gamma(\bigcup_{J_\tau} B_k(s))) \to \lambda \omega(\phi_k(t)) \qquad \text{as} \quad \tau \to 0+ \quad ,$$

where we have applied Proposition 7.2 and the equicontinuity of $\{x_n : n \ge k\}$. Now, we notice that $\phi(t) \equiv 0$ in case $\phi' = \lambda\omega(\phi)$ and $\phi(0) = 0$, since $\rho(t) = \phi(\lambda^{-1}t)$ satisfies $\rho' = \omega(\rho)$ and $\rho(0) = 0$, and therefore $\rho(t) \equiv 0$. Hence, we may proceed as in the proof of Theorem 2.2 .

q.e.d.

2. Galerkin approximations for countable systems

Let us start with the linear problem

(4) $$x_i' = \sum_{j\geq 1} a_{ij}(t)x_j + b_i(t) \quad , \quad x_i(0) = c_i \quad \text{for} \quad i \geq 1 .$$

Theorem 7.3. Let $J = [0,\alpha]$; $a_{ij} \in C(J)$ and $b_i \in C(J)$; $\hat{a}_{ij} = \max_J |a_{ij}(t)|$ and $\hat{b}_i = \max_J |b_i(t)|$; $\exp(\hat{A}t)$ be defined in J . Let c and $\hat{b}$ be in

$$D_\alpha = \{x \in \mathbb{R}^{\mathbb{N}} : \sum_{j\geq 1} e_{ij}^{\hat{A}\alpha}|x_j| < \infty \text{ for each } i \geq 1\} .$$

Let x be that solution of (4) which is the limit of the successive approximations starting with $x^0(t) \equiv c$, and y^N be the solution of

(4_N) $$(y_i^N)' = \sum_{j\leq N} a_{ij}(t)y_j^N + b_i(t) \quad , \quad y_i^N(0) = c_i \quad \text{for} \quad i \leq N.$$

Then $y_i^N(t) \to x_i(t)$ uniformly on J , for each $i \geq 1$.

Proof. By Theorem 6.5 ,

$$\hat{x}(t) = e^{\hat{A}t}|c| + \int_0^t e^{\hat{A}(t-s)}\hat{b}ds$$

is a solution on J of $x' = \hat{A}x + \hat{b}$, $x(0) = |c|$, where $|c| = (|c_1|,|c_2|,..)$ and $\hat{x}$ is the limit of the successive approximations $\hat{x}^n$, defined by $\hat{x}^0(t) \equiv |c|$ and

$$\hat{x}^{n+1}(t) = |c| + \int_0^t \{\hat{A}\hat{x}^n(s) + \hat{b}\}ds \quad \text{for} \quad n \geq 0 .$$

Now, consider the successive approximations

$$x^0(t) \equiv c \quad , \quad x^{n+1}(t) = c + \int_0^t \{A(s)x^n(s) + b(s)\}ds \quad \text{for} \quad n \geq 0 .$$

By induction, it is easy to see that $|x^{n+1}(t) - x^n(t)| \leq \hat{x}^{n+1}(t)-\hat{x}^n(t)$ for $n \geq 0$ and $t \in J$. Clearly, this implies that (x^n) converges to a solution of (4) which we have denoted by x in the theorem , and $|x(t)| \leq \hat{x}(t)$ in J .
Let $\phi_i(t) = |x_i(t) - y_i^N(t)|$ for some $N \geq 1$ and $i \leq N$. Then

(5) $$\phi_i(t) \leq \int_0^t \sum_{j\leq N} \hat{a}_{ij}\phi_j(s)ds + \int_0^t \sum_{j\geq N+1} \hat{a}_{ij}\hat{x}_j(s)ds \text{ for } i \leq N.$$

Let $\hat{A}_N$ be the N×N matrix $(\hat{a}_{ij})$ with $i,j \leq N$. Then (5) implies

$$\phi_i(t) \leq \int_0^t \sum_{j\leq N} e_{ij}^{\hat{A}_N(t-s)} \int_0^s \sum_{k\geq N+1} \hat{a}_{jk}\hat{x}_k(\tau)d\tau ds \quad \text{for} \quad i \leq N .$$

Therefore,

$$\phi_i(t) \le \int_0^t \sum_{j\le N} e_{ij}^{\hat{A}_N(t-s)} [\hat{x}_j(s) - |c_j| - \hat{b}_j s - \int_0^s \sum_{k\le N} \hat{a}_{jk}\hat{x}_k(\tau)d\tau]\,ds \ .$$

Let

$$u(t)=(\hat{x}_1(t),\dots,\hat{x}_N(t)) \text{ and } v(t) = \int_0^t e^{\hat{A}_N(t-s)}(u(s) - \int_0^s \hat{A}_N u(\tau)d\tau)ds.$$

Thus, we have $v' = \hat{A}_N v + u(t) - \int_0^t \hat{A}_N u(\tau)d\tau$ and therefore

$$w' = \hat{A}_N w \quad \text{and} \quad w(0) = 0 \quad \text{for} \quad w(t) = v(t) - \int_0^t u(s)ds \ .$$

Hence, $v(t) = \int_0^t u(s)ds$. This implies

$$\phi_i(t) \le \int_0^t \{\hat{x}_i(s) - \sum_{k\le N} e_{ik}^{\hat{A}_N(t-s)}(|c_k| + \hat{b}_k s)\}ds$$

$$\le \sum_{n\ge 0} \frac{\alpha^{n+1}}{(n+1)!} [\sum_{k\ge 1} \hat{A}^n_{ik}|c_k| - \sum_{k\le N} \hat{A}^n_{Nik}|c_k|]$$

$$+ \sum_{n\ge 0} \frac{\alpha^{n+2}}{(n+2)!} [\sum_{k\ge 1} \hat{A}^n_{ik}\hat{b}_k - \sum_{k\le N} \hat{A}^n_{Nik}\hat{b}_k] \qquad \text{for} \quad i \le N \ .$$

By induction on n it is easy to see that the brackets [...] tend to zero as $N \to \infty$, for each $n \ge 1$. Therefore, we split the sums $\sum_{n\ge 0}$ into $\sum_{n\le m-1} + \sum_{n\ge m}$, and estimating the second ones due to the fact that c and $\hat{b}$ are in D_α , we obtain for $N \to \infty$

$$\overline{\lim_{N\to\infty}} |x_i(t) - y_i^N(t)| \le \frac{\alpha}{m+1} \sum_{k\ge 1} e_{ik}^{\hat{A}\alpha}|c_k| + \frac{\alpha^2}{(m+1)^2} \sum_{k\ge 1} e_{ik}^{\hat{A}\alpha}\hat{b}_k \ .$$

Now, we let $m \to \infty$ to obtain the assertion.

q.e.d.

The following nonlinear version of Theorem 7.3 is an immediate consequence of the proof just given.

Theorem 7.4. Let $J = [0,\alpha]$, $f_i: J\times\mathbb{R}^{\mathbb{N}} \to \mathbb{R}$ be continuous in $t \in J$ and such that

$$|f_i(t,x) - f_i(t,y)| \le \sum_{j\ge 1} a_{ij}|x_j - y_j| \qquad \text{for} \quad i = 1,2,\dots \ ,$$

with constants $a_{ij} \geq 0$ such that $\exp(A\alpha)$ exists. Suppose also that $c \in \mathbb{R}^{\mathbb{N}}$ and $b \in \mathbb{R}^{\mathbb{N}}$, defined by $b_i = \max_J |f_i(t,0)|$, are in D_α (c.p. Theorem 7.3) .

Then system (1) has a solution x that is the uniform limit of the successive approximations starting with $x^0 = c$, the truncated system (1_N) has a unique solution y^N for each $N \geq 1$, and $y_i^N(t) \to x_i(t)$ as $N \to \infty$ uniformly in J , for every $i \geq 1$.

The next theorem shows that Theorem 6.4 , Corollary 6.1 and Corollary 6.2 may be proved by means of Galerkin approximations.

Theorem 7.5. Let $J = [0,\alpha]$, $f_n: J\times\mathbb{R}^{\mathbb{N}} \to \mathbb{R}$ continuous (w.r. to the metric d) and $|f_n(t,x)| \leq \phi_n(t,|x|)$ in $J\times\mathbb{R}^{\mathbb{N}}$, where $\phi_n(t,\rho)$ is monotone increasing in ρ and such that

$$\rho_n' = \phi_n(t,\rho) \quad , \quad \rho_n(0) = |c_n| \qquad \text{for} \quad n \geq 1$$

has a nonnegative solution $\hat{\rho}$ on J .
Then (1_N) has a solution y^N such that $|y_i^N(t)| \leq \hat{\rho}_i(t)$ for $i \leq N$ and $t \in J$, and the sequence (y^N) has a subsequence that converges to a solution of (1) . In case (1) has only one solution x satisfying $|x(t)| \leq \hat{\rho}(t)$ on J , (y^N) converges to x .

Proof. Since ϕ_i is monotone increasing in ρ , we have

$$\phi_i(t,\hat{\rho}_1,\ldots,\hat{\rho}_N,0,\ldots) \leq \hat{\rho}_i' \qquad \text{for} \quad i \leq N \quad .$$

Hence, (1_N) has a solution y^N satisfying

$$|y_i^N(t)| \leq \hat{\rho}_i(t) \qquad \text{and} \qquad |y_i^N(t) - y_i^N(\bar{t})| \leq |\hat{\rho}_i(t) - \hat{\rho}_i(\bar{t})| \quad .$$

Therefore, we find a subsequence (y^m) of (y^N) such that

$$\lim_{m\to\infty} y_i^m(t) = x_i(t)$$

uniformly on J , for some x and each $i \geq 1$. Since f_i is continuous w.r. to d , x is a solution of (1) , and $|x(t)| \leq \hat{\rho}(t)$ on J . Now, the last assertion is evident.

q.e.d.

Convergence of Galerkin approximations is also useful if we want to establish comparison theorems, in particular the nonnegativity of solutions of problem (1) . Evidently, the quasimonotonicity of

$f = (f_1, f_2, \dots)$ in the sense of $\mathbb{R}^n$, i.e. f_i is monotone increasing in the x_j with $j \neq i$, is much easier to check than quasimonotonicity according to Definition 5.2. For example, let $A = (a_{ij})$ be a matrix such that $a_{ij} \geq 0$ for $i \neq j$ and $\exp(|A|t)$ exists for $t \in [0,\rho)$, and let $c = |c| \in D_\alpha$ for some $\alpha < \rho$. Then the solution $x(t) = e^{At}c$ is nonnegative since the Galerkin approximations are nonnegative.

3. Examples

(i) Branching processes. Let us consider again the branching system (cp. Example (i) in 4.4)

$$x_i' = -a_{ii}x_i + \sum_{j \neq i} a_{ij}x_j \quad , \quad x_i(0) = c_i \geq 0 \text{ for } i \geq 1 , \tag{6}$$

where $c \in l^1$ and $|c|_1 = 1$. Let $A = (a_{ij})$ and $A^- = (a_{ij}^-)$ with $a_{ij}^- = a_{ij}$ for $i \neq j$ and $a_{ii}^- = -a_{ii}$. Suppose $\exp(At)$ exists for $t \in [0,\alpha]$ and $e^{A\alpha}c$ is defined. Then $x(t) = e^{A^- t}c$ is a nonnegative solution of (6) on $[0,\alpha]$. Let y^N be the Galerkin approximations. Then $0 \leq y_i^N(t) \leq y_i^{N+1}(t)$ for $i \leq N$ and

$$\begin{aligned} \Big(\sum_{i \leq N} y_i^N(t)\Big)' &= -\sum_{i \leq N}\Big(\sum_{k \neq i} a_{ki}\Big) y_i^N(t) + \sum_{i \leq N} \sum_{1 \leq j \neq i}^{N} a_{ij} y_j^N(t) \\ &= -\sum_{i \geq N+1} \sum_{j \leq N} a_{ij} y_j^N(t) \leq 0 \quad . \end{aligned} \tag{7}$$

Hence,

$$\sum_{i \leq N} y_i^N(t) \leq \sum_{i \leq N} c_i \leq 1 \quad .$$

Since $y_i^N(t) \to x_i(t)$ as $N \to \infty$, we have $|x(t)|_1 \leq 1$ in $[0,\alpha]$. In particular, suppose we have a pure "death" process, i.e. $a_{ij} = 0$ for $i > j$. Then (7) implies $|x(t)|_1 \equiv 1$. In general it is rather difficult to see wether $|x(t)|_1 \equiv 1$ or $|x(t)|_1 < 1$ for some t. The positive result for a pure "death" process may be generalized as follows.

Suppose that there exists an m such that $a_{ij} = 0$ for $i > j+m$, and let

$$\alpha_n = \max\{a_{jj} : j = n-m+1, \dots, n\} \qquad \text{for } n \geq m \quad . \tag{8}$$

Then we have $\sum_{n \geq 1} x_n(t) \equiv 1$ provided $\sum'_{n \geq m} \frac{1}{\alpha_n}$ is divergent, where the

prime indicates summation over n such that $\alpha_n \neq 0$.
To prove this result, let us start with (7)

$$\Big(\sum_{i\le N} y_i^N(t)\Big)' = -\sum_{j\le N}\sum_{i\ge N+1} a_{ij}y_j^N(t) = -\sum_{j=N-m+1}^{N}\Big(\sum_{i\ge N+1} a_{ij}\Big)y_j^N(t)$$

$$\text{(9)} \qquad \ge -\alpha_N \sum_{j=N-m+1}^{N} x_j(t) \quad .$$

Now, suppose $1-\delta \ge \sum_{i\ge 1} x_i(t)$ for some $\delta > 0$ and some $t > 0$. Integrating (9) and keeping in mind that $(y_i^N)_{N\ge i}$ is monotone convergent to $x_i(t)$, we obtain

$$\text{(10)} \quad \int_0^t \sum_{j=N-m+1}^{N} x_j(s)ds \ge \alpha_N^{-1}(\delta - 1 + \sum_{i\le N} c_i) \quad .$$

Since $\sum_1^N c_i \to 1$ as $N \to \infty$, we find $N_0 \ge m$ such that

$$\delta - 1 + \sum_{i\le N} c_i \ge \delta/2 \qquad \text{for} \quad N \ge N_0 \quad .$$

Summing over $N \ge N_0$ in (10), we obtain the contradiction

$$\frac{\delta}{2}\sum_{N\ge N_0}{}' \frac{1}{\alpha_N} \le \sum_{N\ge N_0}\int_0^t \sum_{j=N-m+1}^{N} x_j(s)ds$$

$$\le m\int_0^t \sum_{j\ge 1} x_j(s)ds \le mt < \infty \quad .$$

q.e.d.

Let us show what we can obtain without the assumption that exp(At) be defined in $[0,\alpha]$ and $c \in D(\exp A\alpha)$. We still have $y_i^N(t) \to x_i(t)$ as $N \to \infty$, for every $t \ge 0$, $0 \le x_i(t) \le 1$ and $\sum_{i\ge 1} x_i(t) \le 1$. Furthermore, it is obvious that

$$y_i^N(t) = e^{-a_{ii}t}c_i + \int_0^t e^{-a_{ii}(t-s)} \sum_{j\neq i}^{N} a_{ij}y_j^N(s)ds$$

for $i \le N$

implies

$$x_i(t) = e^{-a_{ii}t}c_i + \int_0^t e^{-a_{ii}(t-s)} \sum_{j\neq i} a_{ij}x_j(s)ds$$

for every $i \ge 1$.

Therefore, x_i is locally absolutely continuous and $x = (x_i)_{i\ge 1}$ satisfies the initial condition $x(0) = c$ and (6) for almost all $t \geq 0$. In particular, Dini's theorem implies $y_i^N(t) \to x_i(t)$ uniformly on every compact interval, but in general we can not assert that $\sum_{j\neq i} a_{ij}x_j(t)$ is continuous, and therefore x may not be a classical solution of (6). Since uniqueness is also a nontrivial problem, let us mention that the solution x just obtained is minimal in the set of all absolutely continuous solutions u satisfying $0 \leq u_i(t) \leq 1$ and $\sum_{i\geq 1} u_i(t) \leq 1$. In fact, let

$$u^N = (u_1,\dots,u_N,0,\dots) \quad .$$

Then

$$u_i^{N\prime} \geq - a_{ii}u_i^N + \sum_{j\neq i} a_{ij}u_j^N \qquad \text{a.e. in} \quad t \geq 0$$

and $u_i^N(0) = c_i$ for $i \leq N$. Therefore $y_i^N(t) \leq u_i^N(t)$ for all $i \leq N$, and this implies $x_i(t) \leq u_i(t)$ in $t \geq 0$, for every $i \geq 1$.

(ii) Degradation of polymers. Let us solve Simha's system (see Remark (vi) in § 6)

$$\text{(11)} \qquad x_i' = -(i-1)\alpha x_i + 2\alpha \sum_{j\geq i+1} x_j \ , \ x_i(0) = c_i \quad \text{for} \quad i \geq 1 \ ,$$

where $\alpha > 0$, $c_i \geq 0$ and $\sum_{i\geq 1} ic_i = K_0$. Consider the Galerkin approximation y^N for (11) . Obviously, $0 \leq y_i^N(t) \leq y_i^{N+1}(t)$ in $[0,\infty)$ for $i \leq N$, and

$$\Big(\sum_{i\leq N} iy_i^N(t)\Big)' = - \alpha \sum_{i=2}^{N} (i-1)iy_i^N(t) + 2\alpha \sum_{i=1}^{N-1} \sum_{j=i+1}^{N} iy_j^N = \alpha\cdot 0 = 0 \quad .$$

Hence,

$$\sum_{i\leq N} iy_i^N(t) = \sum_{i\leq N} ic_i \qquad \text{for every} \quad N \geq 1 \quad .$$

However, $(y_i^N)_{N\geq i}$ is equicontinuous since

$$|y_i^N(t) - y_i^N(\bar{t})| \leq \alpha\Big|\int_t^{\bar{t}}\{-(i-1)y_i^N(s) + 2\sum_{j=i+1}^{N} y_j^N(s)\}ds\Big|$$

$$\leq 3\alpha|t-\bar{t}|\sum_{j\leq N} jc_j \leq 3\alpha K_0|t-\bar{t}| \quad .$$

Therefore, (y_i^N) is uniformly convergent on every bounded subinterval of

$[0,\infty)$, say $y_i^N(t) \to x_i(t)$ as $N \to \infty$. Since (y_i^N) is increasing w.r. to N , and

$$\sum_{i\leq N} i y_i^N(t) = \sum_{i\leq N} i c_i ,$$

we obtain

$$\sum_{i\leq N} i c_i \leq \sum_{i\leq N} i x_i(t) \leq K_o \quad \text{for every } N \geq 1 ,$$

and therefore $\sum_{i\geq 1} i x_i(t) \equiv K_o$ in $[0,\infty)$. In particular, $\sum_{j\geq 1} x_j(t)$ is continuous since the series is uniformly convergent on every bounded interval. Now, it is easy to see that x is a solution of (11) . In fact, we have for $N \geq i$

$$|x_i(t) - c_i - \alpha\int_0^t \{-(i-1)x_i(s) + 2\sum_{j\geq i+1} x_j(s)\}ds|$$

$$\leq x_i(t) - y_i^N(t) + \alpha\int_0^t (i-1)(x_i(s) - y_i^N(s))ds$$

$$+ 2\alpha\int_0^t \sum_{j\geq N+1} x_j(s)ds + 2\alpha\int_0^t \sum_{j=i+1}^{N} (x_j(s) - y_j^N(s))ds .$$

Obviously, the first three terms go to zero as $N \to \infty$, and the same is true for the last term since we have the estimate

$$\sum_{j=i+1}^{N} (x_j(s) - y_j^N(s)) \leq \sum_{j=i+1}^{m} (x_j(s) - y_j^N(s)) + K_o/(m+1) \text{ for } N > m.$$

Thus, we have found a solution x of (11) on $[0,\infty)$ which satisfies the side condition $\sum_{i\geq 1} i x_i(t) \equiv K_o$.

Let us note that the matrix $A = (a_{ij})$ corresponding to (11) is upper-diagonal and satisfies $\sum_{i\geq 1} |a_{ij}| = 3\alpha(j-1)$ for every $j \geq 1$. Hence, by [...] in Example 6.8 (d) with $N = 0$, $\exp(|A|t)$ exists in $[0,\infty)$ and $\sum_{i\geq 1} e_{ij}^{|A|t} < \infty$ for every $j \geq 1$. In order to apply Theorem 6.5 , we should have

$$\sum_{j\geq 1} e_{ij}^{|A|t} c_j = \sum_{j\geq 1} j^{-1} e_{ij}^{|A|t}(jc_j) < \infty \quad \text{for every } i \geq 1 ,$$

i.e. $\sup\{j^{-1}e_{ij}^{|A|t} : j \geq 1\} < \infty$ for each i . But a simple calculation yields $|A|_{ij}^2 = 2\alpha^2(3j-i-4)$ for $j > i$ and therefore

$$|A|_{ij}^3 \geq |A|_{ij}^2 |a_{jj}| = 2\alpha^3(3j-i-4)(j-1) .$$

Hence, $j^{-1}e_{ij}^{|A|t} \to \infty$ for each $t > 0$ and $i \geq 1$, and consequently we find c such that $\sum_{i\geq 1} ic_i < \infty$, but $\sum_{j\geq 1} e_{ij}^{|\bar{A}|t}c_j = \infty$ for every $i \geq 1$. Therefore, this example shows in particular that the Galerkin approximations may converge though $\exp(|A|t)|c|$ does not exist for $t > 0$.

4. Remarks

(i) Theorem 7.1 is taken from Deimling [48]. The hypotheses of this theorem are not sufficient for convergence of successive approximations. A well known counter example is as follows:

$X = \mathbb{R}^1$, $c = 0$ and

$$f(t,x) = \begin{cases} 0 & \text{for } t = 0 \text{ , } x \in \mathbb{R}^1 \\ 2t & \text{for } t \in (0,1] \text{ and } x < 0 \\ 2t-4x/t & \text{for } t \in (0,1] \text{ and } 0 \leq x \leq t^2 \\ -2t & \text{for } t \in (0,1] \text{ and } t^2 < x \end{cases} .$$

It is easy to see that f is continuous, bounded by $M = 2$ and monotone decreasing in x, i.e. (3) is satisfied with $\omega(t,\rho) \equiv 0$. The successive approximations starting with $x_0(t) \equiv 0$ do not converge, see Coddington/Levinson [34].

Suppose that f satisfies the stronger condition

$$|f(t,x) - f(t,y)| \leq \omega(t,|x-y|)$$

where ω is of class U_1. Then it is easy to prove convergence of the successive approximations provided $\omega(t,\rho)$ is monotone increasing in ρ; see Deimling [48]. In the same paper a proof of convergence has been given in case ω is not increasing. Unfortunately, this proof is not correct. Evans/Feroe [61] have a counterexample for $X = \mathbb{R}^1$ and a proof for $X = \mathbb{R}^n$ with $n \geq 2$. The infinite dimensional problem is still open.

(ii) Convergence of Galerkin approximations for countable systems has been studied by several authors. Let us consider some examples.

Shaw [159] has $c \in l^1$, a constant matrix (a_{ij}) such that $\sup_i \sum_{j\geq 1} |a_{ij}| < \infty$ and $b\colon J \to l^1$ continuous. Obviously, the conditions of Theorem 7.3 are satisfied.

McClure/Wong [122] are concerned with $x' = Ax$, $x(0) = c \in l^1$ where $\sup_i \operatorname{Re} a_{ii} < \infty$ and $\sup_j \sum_{i\neq j} |a_{ij}| = M < \infty$; see Remark (vii) to § 6. They prove convergence provided A satisfies in addition

$$|a_{jj}| \geq \delta + \sum_{i \neq j} |a_{ij}| \quad \text{for some} \quad \delta > 0 \quad \text{and every } j \geq 1;$$

obviously this condition implies that A has a bounded inverse on l^1. However, it is trivial to prove convergence without this extra condition. In fact, consider the equivalent system of integral equations, let x be the unique l^1-solution, y^N the N-th approximation, and $\omega = \sup_i \operatorname{Re} a_{ii}$. Then

$$\psi_N(t) = \sum_{i \leq N} |x_i(t) - y_i^N(t)| \leq M \int_0^t e^{\omega(t-s)} \psi_N(s) ds$$

$$+ M \int_0^t e^{\omega(t-s)} \sum_{j \geq N+1} |x_j(s)| ds$$

which implies $\psi_N(t) \to 0$ uniformly on bounded subintervals of $[0,\infty)$.

Moszynski/Pokrzywa [126] have shown convergence if there is a solution x of (1) such that $\hat{x} \in D_\alpha$, where $\hat{x}_i = \max_{[0,\alpha]} |x_i(t)|$; in general this is the case if $\exp(2\hat{A}\alpha)$ exists and $|c|$ and $\hat{b}$ are in $D_{2\alpha}$. Obviously, this result is a trivial special case of Theorem 7.4 . They have also proved convergence of successive approximations for (1) , under conditions stronger than those in Theorem 7.4 .

Bellman [8] considered $x' = A(t)x$, $x(0) = c$ with $c \in l^p$ for some $p \geq 1$, a_{ij} measurable , $\sum_{i,j \geq 1} |a_{ij}(t)| = \phi(t) < \infty$ for $p = 1$ and

$$\Big[\sum_{i \geq 1} \Big(\sum_{j \geq 1} |a_{ij}(t)|^q\Big)^{p/q}\Big]^{1/p} = \phi(t) < \infty \quad \text{in case } p > 1 ,$$

where $\phi \in L^1(J)$ and $p^{-1} + q^{-1} = 1$. By means of successive approximations, he proved that there exists a unique "solution" $x: J \to l^p$. This has also been done much earlier by Reid [154] . He has also shown that the Galerkin approximations converge to x .

Shaw [160] has p = 1 and measurable a_{ij} such that

$$(12) \qquad \sup_i \sum_{j \geq 1} |a_{ij}(t)| \leq M \quad \text{and} \quad \sup_j \sum_{i \geq 1} |a_{ij}(t)| \leq M \quad \text{in } J .$$

Clearly, under such "Carathéodory"-conditions the "solutions" and the Galerkin approximations are only absolutely continuous and the differential equation is only satisfied almost everywhere. These results of Bellman and Shaw do not follow directly from our theorems. However, they are very easy to prove by means of the corresponding system of integral equations. For example, let p = 1 and $\sup_j \sum_{i \geq 1} |a_{ij}(t)| \leq \phi(t)$

with $\phi \in L^1(J)$, a condition much weaker than (12) . Then it is obvious that the system

$$x_i(t) = c_i + \int_0^t \sum_{j\geq 1} a_{ij}(s)x_j(s)ds \quad \text{for} \quad i \geq 1$$

has a unique continuous solution $x\colon J \to l^1$, and we have

$$\psi_N(t) = \sum_{i\leq N} |x_i(t) - y_i^N(t)| \leq \int_0^t \phi(s)\psi_N(s)ds + \int_0^t \phi(s) \sum_{j\geq N+1} |x_j(s)|ds$$

which clearly implies $\psi_N(t) \to 0$ as $N \to \infty$, uniformly on J .
Žautykov [198] considered the nonlinear system (1) in $X = l^\infty$, where $f_n\colon J\times\overline{K}_r(0) \to \mathbb{R}^1$ is assumed to be continuous in t and to satisfy the so called "strengthened Cauchy-Lipschitz condition" introduced before by Persidskii [137] , i.e.

$$(13) \qquad |f_n(t,x_1,\ldots,x_m,x_{m+1},\ldots) - f_n(t,x_1,\ldots,\overline{x}_m,\overline{x}_{m+1},\ldots)| \leq \varepsilon_m \sup_{j\geq m} |x_j-\overline{x}_j|$$

for $m \geq 1$, x and $(x_1,\ldots,x_{m-1},\overline{x}_m,\ldots)$ in $\overline{K}_r(0) \subset l^\infty$ and $\varepsilon_m \to 0$ as $m \to \infty$. Moreover, he has $|f_n(t,0)| \leq M$ for every $n \geq 1$ and $t \in J$. Let ε_1 in (13) be independent of n . Then we may extend f_n to $J\times\mathbb{R}^{\mathbb{N}}$ such that the Lipschitz constant ε_1 does not change, and such that $|f_n(t,x)| \leq M+\varepsilon_1 r$ on $J\times\mathbb{R}^{\mathbb{N}}$. Now, Theorem 7.4 applies, and in case $|c|_\infty < r$ we thus find an interval $[0,\alpha] \subset J$ such that the Galerkin approximations of the original problem tend to the unique solution on $[0,\alpha]$. The same approach applies in case ε_m is independent of n for some $m \geq 2$.

(iii) Let us consider some countable systems that arise in connection with Fourier's method.
An early paper is that of Lewis [106] who studied the boundary value problem for the heat equation, i.e. Example 1 from the introduction, and the corresponding problem for the nonlinear wave equation. For the parabolic problem, the countable system is

$$x_n' + n^2x_n = f_n(t,x_1,x_2,\ldots) \ , \quad x_n(0) = c_n \ .$$

Lewis assumes that $\sum_{n\geq 1} c_n^{\,2}n^4 < \infty$ which means that the initial condition $u(x,0) = \phi(x)$ satisfies $\phi'' \in L^2(0,\pi)$. Therefore, he considers the set

$D = \{x \in l^2 : \sum_{n \geq 1} n^4 x_n{}^2 \leq r^2\}$ for some $r > 0$, $f = (f_n)$ satisfying $\sum_{n \geq 1} n^2 f_n{}^2 \leq M$ and

$$|f(t,x) - f(t,\bar{x})|_2 \leq L\Big(\sum_{n \geq 1} n^2 (x_n - \bar{x}_n)^2\Big)^{1/2} \quad \text{in } J \times D .$$

By means of successive approximations he obtains a local solution. Under similar conditions, he proves convergence of the Galerkin approximations for the hyperbolic problem. There is also a rather complete bibliography of some earlier papers on countable systems not mentioned in these notes.

To mention one more recent interesting paper at least, consider the mixed problem

$$u_{tt} - \Big(a_o + a_1 \int_0^L u_x^2(x,t)dx\Big)u_{xx} = 0 \quad (\text{with } a_o \geq 0, a_1 > 0)$$

$$u(0,t) = u(L,t) = 0 , \; u(x,0) = \phi(x) \text{ and } u_t(x,0) = \psi(x) \quad \text{in } [0,L],$$

which describes the small amplitude vibration of a string in which the dependence of the tension on the deformation cannot be neglected. Let

$$\phi(x) = \sum_{k \geq 1} \alpha_k \sin(k\pi x/L) \quad , \quad \psi(x) = \sum_{k \geq 1} \beta_k \sin(k\pi x/L) \quad .$$

In attempting to find u as the Fourier series $u(x,t) = \sum_{k \geq 1} u_k(t) \sin\frac{k\pi x}{L}$, one is led to study the countable system

$$(14) \quad u_i'' + i^2\Big(\bar{a}_o + \bar{a}_1 \sum_{j \geq 1} j^2 u_j^2\Big)u_i = 0 , \; u_i(0) = \alpha_i , \; u_i'(0) = \beta_i$$

$$\text{for } i \geq 1 , \text{ where } \bar{a}_o = a_o \pi^2 L^{-2} \text{ and } \bar{a}_1 = a_1 \pi^4/(2L^3) \quad .$$

By means of Galerkin approximation, Dickey [53] proved local existence of a C^2-solution to (14) provided

$$\sum_{k \geq 1} k^4 \alpha_k^2 < \infty \quad , \quad \sum_{k \geq 1} k^2 \beta_k^2 < \infty \text{ and } a_o + a_1 \sum_{k \geq 1} k^2 \alpha_k^2 \neq 0 \quad .$$

It seems to be open whether such a classical solution exists for all $t \geq 0$; see also Dickey [54] .

Clearly, the character of the countable system for the Fourier coefficients of the solution depends very much on the choice of the orthonormal system. While this choice is obvious in the boundary value problems above, it is sometimes a crucial point for the application of Fourier's method to find such an appropriate system of functions of the space variables. An example of much influence is Hopf [79] who established the existence of a weak solution for all time of the Navier -

Stokes equations in a bounded region. See also Dolph/Lewis [56] who used Fourier's method to prove instability of plane Poiseuille flow and Rautmann [146] for more recent references in this direction. Another example is Chalon/Shaw [31] who discussed the initial value problem

$$x'' + \beta x' + \alpha x = 0 \quad , \quad x(0) = x_o \quad , \; x'(0) = x_1$$

where α is a uniformly distributed uncertain parameter varying in some interval $J = [\alpha_1, \alpha_2]$. They consider expansions

$$x(t) \quad = \quad \sum_{i \geq 1} x_i(t)\phi_i(\alpha) \quad ,$$

where $\{\phi_i\}$ is either the trigonometric system or the system of Walsh functions, see e.g. Fine [64] , Roider [157] . It turns out that the countable system for the coefficients x_i with respect to the Walsh functions is so simple that the results of Shaw in Remark (ii) apply, while it becomes more difficult for the trigonometric system.

(iv) Let us emphasize that Galerkin's method is obviously not the only possibility to truncate a countable system. Sometimes one is interested in truncations which preserve certain properties of solutions of the original system. While Galerkin's method turned out to be optimal for Example (ii) in sec. 3 , where the solutions of the truncated system have all properties wanted for the solution of the original system, this is not the case in Example (i) of sec. 3 since $\sum_{i \leq N} y_i^N(t) < 1$ in general. This defect may also be observed in Example 1.3 , where truncation should preserve moment properties like $u_n(t) \geq 0$, $u_1{}^2(t) \leq u_2(t)$, etc. ; see Bellman/Wilcox [20] and the references given there.

(v) We have already mentioned in the introduction that most of the very early papers on countable systems are concerned with analytic solutions of systems with analytic right hand sides. Perhaps the most general of these results are contained in a remarkable paper of Wintner [191] . Consider the formal power series

$$\phi(x) \quad = \quad \sum_{n \geq o} \; \sum_{i_1 \geq 1} \cdots \sum_{i_n \geq 1} a^{(n)}_{i_1 \ldots i_n} x_{i_1} \cdots x_{i_n}$$

in $x \in \mathbb{R}^{\mathbb{N}}$ or $x \in \mathbb{C}^{\mathbb{N}}$, where the coefficients are assumed to be invariant under permutations of $i_1, \ldots, i_n$, without loss of generality. In [191] , ϕ is said to be analytic ("regular") in $K_r(0) \subset l^2$ if

$$\phi^N(x) \quad = \quad \phi(x_1, \ldots, x_N, 0, 0, \ldots)$$

is analytic in $K_r(0) \cap \mathbb{R}^N$ (or $\cap\ \mathbb{C}^N$) for every $N \geq 1$, $\lim_{N\to\infty} \phi^N(x) = \phi(x)$ in $K_r(0)$, and to each $\varepsilon \in (0,r)$ there is a constant $M_{r-\varepsilon}$ such that $|\phi^N(x)| \leq M_{r-\varepsilon}$ in $K_{r-\varepsilon}(0) \cap \mathbb{R}^N$ (or $\cap\ \mathbb{C}^N$) for every $N \geq 1$. Notice that the analyticity of ϕ in $K_r(0)$ does not imply uniform convergence in any smaller ball ; $\phi(x) = \sum x_i^2$ is a simple counterexample. Since there are bounded bilinear forms which are not absolutely bounded, the analyticity of ϕ in $K_r(0)$ does also not imply that $\tilde{\phi}$, defined by taking $|a^{(n)}_{i_1 \ldots i_n}|$ in place of $a^{(n)}_{i_1 \ldots i_n}$ in the series of ϕ , is analytic in any ball $K_\rho(0)$; consider for example

$$\phi(x) = \sum_{i\geq 1} \sum_{j\geq 1}{}' \frac{1}{i-j} x_i x_j ,$$

where the prime indicates $j \neq i$; see Riesz [156,p.155] . Nevertheless, the main result of Wintner is as follows:
Suppose that every f_i is analytic in $\Omega = K_\delta(0) \times K_r(0)$, with $K_\delta(0) \subset \mathbb{R}$ (or $\mathbb{C}$) and $K_r(0) \subset l^2$, and $\sum_{i\geq 1} |f_i(t,x)|^2 \leq M^2$ in Ω . Then the initial value problem $x_i' = f_i(t,x)$, $x_i(0) = 0$ for $i \geq 1$ has an analytic solution in $K_\rho(0)$, where $\rho = \min\{\delta, r/(2M)\}$. An immediate consequence of this result is the following implicite function theorem which has been applied to some problems of celestial mechanics by Wintner [192] :
Suppose that the functions f_i satisfy the conditions just mentioned ; then the system $x_i = tf(t,x_1,x_2,\ldots)$ for $i \geq 1$ has an analytic solution $x(t)$ in $\{t : |t| < \rho\}$, for some $\rho > 0$.

(vi) Arley/Borchsenius [4] and Hille [75] have some interesting examples showing the pathological behavior of linear countable systems with respect to analyticity of solutions. Even in the case of constant coefficients it may happen that there is a unique solution x , but x is not analytic at $t = 0$. It is also rather easy to construct a function x which is analytic in $\mathbb{C}$ and a matrix A which is analytic in $\mathbb{C}$ with the exception of simple poles such that $x' = A(z)x$ holds in a certain open subset G of $\mathbb{C}$ only, but the poles of A are not on ∂G . Hence, x is not a solution in limit points of points in which it is a solution. Clearly, both phenomenons do not occur in finite linear systems.
On the other hand, solutions of a nonlinear equation with entire right hand side may have singularities. Consider for example

$$(15) \qquad u' = -u + u^2 , \qquad u(0) = \alpha \neq 0 .$$

The solution is given by $u(z) = e^{-z}(\frac{1}{\alpha} - 1 + e^{-z})^{-1}$ and therefore it has singularities. Now, let $x_n(z) = u^n(z)$. By means of (15) we obtain the countable system

(16) $$x_n' = -nx_n + nx_{n+1} \quad , \quad x_n(0) = \alpha^n \quad \text{for} \quad n \geq 1 \quad .$$

Thus we have explained why solutions of countable systems with constant coefficients may have singularities.
Obviously, the transformation of (15) into (16) is also possible for any finite nonlinear system with analytic right hand sides. Apparently, this method is due to Carleman [26] , following an idea of Poincaré who proposed to "solve" nonlinear differential equations by means of linear integral equations in 1908 . It is also mentioned in Bellman [8], who wanted to prove boundedness of solutions of a single nonlinear equation by means of boundedness of solutions to the corresponding infinite linear system. Following Levinson [105] , he has the criterion "Let $c \in l^1$, $\lambda_i < 0$ for each $i \geq 1$ and $\sum_{j \geq i+1} |a_{ij}||c_j| \leq \alpha |c_i|$ for some $\alpha \in (0,1)$ and each $i \geq 1$. Then every solution of

$$x_i' = \lambda_i x_i + \lambda_i \sum_{j \geq i+1} a_{ij} x_j \quad , \quad x_i(0) = c_i \quad \text{for} \quad i \geq 1$$

satisfies $|x_i(t)| \leq (1-\alpha)^{-1}|c_i|$ for $t \geq 0$ and $i \geq 1$".
As an "application", he claims that the solutions of $u' = u + u^2$ with $\alpha = |u(0)| < 1$ are bounded. However, the corresponding system

$$x_n' = nx_n + nx_{n+1} \quad , \quad x_n(0) = u^n(0) \quad \text{for} \quad n \geq 1$$

does not satisfy the conditions of his criterion, and in fact

$$x_1(t) \quad = \quad e^t(\alpha^{-1} + 1 - e^t)^{-1}$$

is unbounded.

(vii) To conclude this chapter, let us give some remarks on countable systems coming from branching processes. Some simple facts and examples may be found e.g. in the books of Bellman [9] , Cox/Miller [36] , Feller [62] , Karlin [81] and Ludwig [110] . Apparently, the most serious early discussion of such systems is contained in Feller [63] and in the remarkable thesis of Arley [3] , who has an interesting application to the theorem of cosmic radiation and some criteria for $\sum_{i \geq 1} x_i(t) \equiv 1$. For a detailed discussion by means of semigroup theory (cp. § 8.3) we refer to the more recent papers of Kato [83] and Reuter [155] .

§ 8 Related Topics

In this final chapter we shall review some further topics hardly mentioned before but closely related to material presented in earlier chapters. It should be as useful to students stopping hereafter their studies in this direction as to those still looking for more sophisticated problems.

1. Carathéodory Conditions.

Let X be a Banach space, $J = [0,a] \subset \mathbb{R}$, $D = \bar{K}_r(x_0) \subset X$ and $f: J \times D \to X$, and consider the initial value problem

$$(1) \qquad x' = f(t,x) \quad , \quad x(0) = x_0 \quad .$$

In the main, we have considered local existence of solutions to (1) in case f is continuous at least. In finite dimensions, however, it is well known that (1) has an absolutely continuous (a.c.) solution if f satisfies Carathéodory's conditions only, i.e. f is Lebesgue measurable in t , continuous in x and such that $|f(t,x)| \leq M(t)$ for some function $M \in L^1(J)$. Since the main definitions cary over to infinite dimensions it is at hand to study the same existence problem in arbitrary Banach spaces.

Let us recall in particular that a function $x: J \to X$ is said to be a.c. (with respect to Lebesgue measure) if to every $\varepsilon > 0$ there exists $\delta > 0$ such that

$$\sum_i |x(\bar{t}_i) - x(t_i)| \leq \varepsilon$$

for every finite family of nonoverlapping subintervals $[t_i,\bar{t}_i] \subset J$ such that

$$\sum_i |\bar{t}_i - t_i| \leq \delta \quad .$$

In case $x(t) = \int_0^t y(s)ds$ with some Bochner integrable function $y: J \to X$, it is easy to see that x is a.c. , differentiable almost everywhere (a.e.) with $x'(t) = y(t)$ a.e. ; the definition of Bochner's integral may be found e.g. in Yosida [197,p.132] , Hille/Phillips [77] . The converse, i.e. "if x is a.c. then x' exists a.e. and

$$x(t) = x(0) + \int_0^t x'(s)ds \quad "$$

is known to be true in finite dimensions and in every reflexive X , see e.g. Komura [89] , but not in every X as may be seen in the following simple

Example 8.1. Let $X = (c_o)$, $J = [0,1]$ and $x: J \to X$ be defined by

$$x_n(t) = \frac{1}{n} \sin(nt) \quad \text{for} \quad n \geq 1 \text{ , } t \in J \text{ .}$$

The function x is a.c. in J since $|x(t) - x(\bar{t})| \leq |t-\bar{t}|$, but x is nowhere differentiable since $(\cos(nt)) \notin (c_o)$ for $t \in [0,1]$.

One natural way to prove local existence for (1) under Carathéodory conditions consists in the approximation of f by continuous functions f_n such that

$$\int_J |f_n(t,x) - f(t,x)|dt \to 0 \quad \text{as} \quad n \to \infty \text{ , for every } x \in D.$$

Then it should be possible to find a solution x_n of (1) , with f_n instead of f , and finally one would like to extract a subsequence of (x_n) convergent to a solution of (1) . However, since we need extra conditions in x , e.g. condition 2.9 or 3.3 , the difficulty consists in the fact that these extra conditions may not be preserved if we replace f by f_n .

We could also look for conditions on f such that the corresponding integral equation

$$(2) \qquad x(t) = x_o + \int_0^t f(s,x(s))ds$$

has a continuous solution, either via fixed point theorems or via approximate solutions to (2) . Let us indicate this last method in the proof of

Theorem 8.1. Let X be a Banach space, $J = [0,a] \subset \mathbb{R}$, $D = \bar{K}_r(x_o) \subset X$; $f: J \times D \to X$ measurable in t , uniformly continuous in x ; $|f(t,x)| \leq M(t)$ with $M \in L^1(J)$. Let

$$(3) \qquad (f(t,x)-f(t,y),x-y)_- \leq \omega(t,|x-y|)|x-y| \quad \text{for } x,y \in D \text{ and almost all } t \in J \text{ ,}$$

where $\omega: J\times\mathbb{R}_+ \to \mathbb{R}_+$ satisfies Carathéodory's conditions, $\omega(t,\cdot)$ is increasing and $\rho(t) \equiv 0$ is the unique solution of the problem $\rho' = \omega(t,\rho)$ a.e. in J , $\rho(0) = 0$.
Then problem (1) has an a.c. solution on $[0,b]$ for every $b \leq a$ such that $\int_0^b M(s)ds \leq r$.

Proof. We consider the Peano approximations

$$(4) \qquad x_n(t) = \begin{cases} x_0 & \text{for } t \leq 0 \\ x_0 + \int_0^t f(s,x_n(s - \frac{1}{n}))ds & \text{for } t > 0 \end{cases}$$

on $[0,b]$. Since $|f(t,x)| \leq M(t)$, the sequence (x_n) is equicontinuous, x_n is a.c. with

$$x_n'(t) = f(t,x_n(t - \frac{1}{n}))$$

a.e. in $[0,b]$ and the function $\varphi(t) = |x_n(t) - x_m(t)|$ is a.c. with $\varphi(0) = 0$ and

$$\varphi(t)\varphi'(t) \leq \omega(t,\varphi(t))\varphi(t) + |y_n(t)-y_m(t)|\varphi(t)$$

a.e. in $[0,b]$, where $y_i(t) = f(t,x_i(t)) - f(t,x_i(t - \frac{1}{i}))$ for $i = n,m$. Since $\rho(t) \equiv 0$ may be approximated by solutions of $\rho' = \omega(t,\rho)+\delta$, $\rho(0) = \delta$, uniformly in $[0,b]$ as $\delta \to 0+$, we only have to show that

$$(5) \qquad \int_0^b |y_n(s) - y_m(s)|ds \to 0 \text{ as } n,m \to \infty .$$

Since $f(t,\cdot)$ is uniformly continuous and

$$\max_{[0,b]} |x_i(t) - x_i(t - \frac{1}{i})| \to 0 \quad \text{as} \quad i \to \infty$$

we have $y_i(t) \to 0$ (a.e.) . Therefore, (5) is true since $|y_i(t)| \leq 2M(t)$.

q.e.d.

An existence theorem, where (3) is replaced by a condition using the measure of noncompactness, is given in Pianigiani [143] ; see also Knight [87] , Pulvirenti [144] . For related theorems on countable systems see the references given in § 7 .

2. Weaker continuity

Let X be a Banach space and let "$\rightharpoonup$" denote weak convergence. Let $J = [0,a] \subset \mathbb{R}$ and $D \subset X$. The map $f: J \times D \to X$ is said to be demicontinuous if $t_n \to t_o$ and $x_n \to x_o \in D$ imply $f(t_n,x_n) \rightharpoonup f(t_o,x_o)$. Several authors have been concerned with existence of solutions to problem (1) in case f is demicontinuous. A (local) solution is understood to be a continuous weakly continuously differentiable function x such that (1) holds on some interval $[0,\delta] \subset J$, where the function x is said to be weakly continuously differentiable in J if there exists a weakly continuous function x' such that $h^{-1}(x(t+h)-x(t)) \rightharpoonup x'(t)$ as $h \to 0$. For example, let us prove the following theorem.

Theorem 8.2. Let X be a Banach space such that X^* is uniformly convex; $D = \overline{K}_r(x_o) \subset X$; $J = [0,a] \subset \mathbb{R}$; $f: J \times D \to X$ demicontinuous, $|f(t,x)| \leq c$ on $J \times D$. Let f satisfy the estimate (3) for $t \in J$ and $x,y \in D$ with ω as in Theorem 8.1. Then problem (1) has a unique solution on $[0,b]$, where $b = \min\{a, r/c\}$.

Proof. We consider the Peano approximations x_n defined by (4) on $[0,b]$. Obviously,

$$(6) \qquad |x_n(t)-x_n(\overline{t})| \leq c|t-\overline{t}| \quad \text{for} \quad n \geq 1 \text{ , } t \text{ and } \overline{t} \in [0,b] .$$

The weak derivative of x_n is given by $x_n'(t) = f(t,x_n(t - \frac{1}{n}))$. We remember (v) of Lemma 3.2, and we notice that (vi) of Lemma 3.2 remains true for weakly differentiable functions. In particular, we have the estimate

$$|Fx - F\overline{x}| \leq d(|x-\overline{x}|) \quad \text{for} \quad x,\overline{x} \in \overline{K}_r(x_o) ,$$

where F denotes the duality map of X and $d: \mathbb{R}_+ \to \mathbb{R}_+$ is some increasing function with $d(\rho) \to 0$ as $\rho \to 0+$. Hence, the function $\varphi(t) = |x_n(t) - x_m(t)|$ is a.c. with $\varphi(0) = 0$ and

$$\varphi(t)\varphi'(t) \leq \omega(t,\varphi(t) + c(\tfrac{1}{n} + \tfrac{1}{m}))(\varphi(t) + c(\tfrac{1}{n} + \tfrac{1}{m})) + $$
$$+ 2Md(c(\tfrac{1}{n} + \tfrac{1}{m})) \quad \text{a.e. in } [0,b] .$$

Let $\psi(t) = \varphi(t) + c(\frac{1}{n} + \frac{1}{m})$. Then

$$\psi^2(t) \leq \alpha_{nm} + 2\int_0^t \omega(s,\psi(s))\psi(s)ds \quad \text{with } \alpha_{nm} \to 0 \text{ as } n,m \to \infty .$$

Given $\varepsilon > 0$ we choose $\delta > 0$ and a continuous function ρ such that

$$\varepsilon \geq \rho(t) = \delta + \int_0^t \omega(s,\rho(s))ds \quad \text{in} \quad [0,b] \quad .$$

As long as $\psi(t) < \rho(t)$, we therefore have

$$\psi^2(t) \leq \alpha_{nm} + 2\int_0^t \omega(s,\rho(s))\rho(s)ds = \alpha_{nm} + \rho^2(t) - \delta^2 \ .$$

Hence, we may choose $n_0 = n_0(\varepsilon)$ such that $\alpha_{nm} \leq \delta^2$ for $n,m \geq n_0$, to obtain

$$\varphi(t) \leq \psi(t) \leq \rho(t) \leq \varepsilon \quad \text{in} \quad [0,b] \quad \text{for} \quad n,m \geq n_0 \quad .$$

Therefore, $x_n(t) \to x(t)$ for some continuous function x , uniformly on $[0,b]$. Now, (4) implies

$$x(t) = x_0 + \int_0^t f(s,x(s))ds \quad ,$$

i.e. x is a solution of (1) . Uniqueness is obvious since $\varphi(t)\varphi'(t) \leq \omega(t,\varphi(t))\varphi(t)$ a.e. and $\varphi(0) = 0$ imply $\varphi \equiv 0$.

q.e.d.

In case $\omega = 0$, Theorem 8.2 is Theorem 9.9 in Browder [21] . Theorems like this one may be used to prove properties of demicontinuous maps like surjectivity etc. along the lines of § 3.6 ; see e.g. Browder [21, Chap. 10] , Crandall [37] . Notice that the condition $|f(t,x)| \leq c$ is satisfied provided r is sufficiently small.
An other type of continuity is the weak continuity of f , i.e. $t_n \to t_0$ and $x_n \rightharpoonup x_0 \in D$ imply $f(t_n,x_n) \rightharpoonup f(t_0,x_0)$. Clearly, a weakly continuous f is also demicontinuous. By a local solution of (1) with weakly continuous f we understand a weakly continuously differentiable function $x: [0,\delta] \to D$, for some $\delta \in [0,a]$, that satisfies (1) on $[0,\delta]$.

<u>Theorem 8.3</u>. Let X be a reflexive Banach space, $D = \bar{K}_r(x_0) \subset X$, $J = [0,a] \subset \mathbb{R}$, $f: J \times D \to X$ weakly continuous and $|f(t,x)| \leq c$ on $J \times D$. Then (1) has a solution x on $[0,b]$, with $b = \min\{a,r/c\}$, such that x is also absolutely continuous in $[0,b]$.

<u>Proof</u>. Consider the Peano approximations x_n from (4) that satisfy (6) and $|x_n(t)| \leq r$ on $[0,b]$. Since a reflexive Banach space is sequentially weakly complete (see [197, p. 124]) there exists a subsequence $(x_m) \subset (x_n)$ and a function x such that $x_m(t) \rightharpoonup x(t)$ on $[0,b]$. For $x^* \in F(x(t) - x(\bar{t}))$, we therefore have

$$|x(t)-x(\bar{t})|^2 = \lim_{m\to\infty} <x_m(t)-x_m(\bar{t}),x^*> \leq c|t-\bar{t}||x(t)-x(\bar{t})| \quad ,$$

i.e. x is Lipschitz and therefore absolutely continuous on $[0,b]$. Since

$$<f(t,x_m(t-\tfrac{1}{m})) - f(t,x(t)),x^*> \to 0 \quad \text{as} \quad m \to \infty$$

for every $x^* \in X^*$, uniformly on $[0,b]$, we obtain

$$(7) \qquad x(t) = x_0 + \int_0^t f(s,x(s))ds \quad \text{in} \quad [0,b] \quad .$$

Hence, x is weakly continuously differentiable with $x'(t) = f(t,x(t))$ in $[0,b]$ (weak derivative) and $x'(t) = f(t,x(t))$ a.e. in $[0,b]$ (strong derivative).

q.e.d.

Results of this type may be found e.g. in Fitzgibbon [65] , Chow/Schuur [32] . Again, $|f(t,x)| \leq c$ is true in $J\times D$ provided r is sufficiently small. By means of these local theorems it is at hand to establish global theorems as in the continuous case. Since f in Theorem 8.1 is compact with respect to the weak topology one could also apply Tychonov's fixedpoint theorem to the integral equation (7) in order to obtain a solution. One could also generalize this result by means of estimates of f in terms of measures of weak noncompactness ; see e.g. Danes [41] , DeBlasi [44] .
In this section we have indicated what everybody familiar with the preceding chapters and the usual tricks available in reflexive or even more special spaces expects: If we weaken the continuity conditions on f then the solutions become weaker and we have to assume more about the space in order to find such solutions.

3. Evolution equations

In § 1.4 we have already refered to the present chapter for some remarks on problem (1) where f is neither bounded nor everywhere defined in a neighborhood of x_0 . Such equations have been called evolution equations by several authors. Let us start with the simple linear

Example 8.2. Let $X = l^1$ and consider the countable system $x_i' = -\alpha_i x_i$, $x_i(0) = x_{0i}$ for $i \geq 1$, where $\alpha_i > 0$ and $\alpha_i \to \infty$ as $i \to \infty$. The corresponding matrix defines an operator A with domain

$$D(A) = \left\{ x \in l^1 : \sum_{i \geq 1} \alpha_i |x_i| < \infty \right\}$$

and range in X . Thus we have to solve

(8) $$x' = Ax \quad , \quad x(0) = x_0 \in D(A)$$

with an unbounded operator A . However, A is densely defined and A is dissipative on D(A) , i.e. $(Ax,x)_- \leq 0$ for $x \in D(A)$, as may be seen from part (iii) of Example 3.1 . Furthermore, $A-\lambda I$ is surjective with $(A-\lambda I)^{-1}: X \to D(A)$ bounded by $|(A-\lambda I)^{-1}| \leq 1/\lambda$ for every $\lambda > 0$. It turns out that these properties of A form a powerful substitute for the violated boundedness in solving (8) in any Banach space. Clearly, our very simple system may be solved explicitely:

For $x_0 \in D(A)$ the function $x: [0,\infty) \to D(A)$, given by $x_i(t) = x_{oi}e^{-\alpha_i t}$ for $i \geq 1$ is the unique continuously differentiable solution of (8) . In fact, we have

$$|x(t+h)-x(t)-hAx(t)| = \sum_{i\geq 1} |x_{oi}| e^{-\alpha_i t} |e^{-\alpha_i h} - 1 + \alpha_i h|$$

$$\leq |h| \sum_{i\leq N} |x_{oi}| \alpha_i (e^{\alpha_i |h|} - 1) + 2|h| \sum_{i\geq N+1} |x_{oi}| \alpha_i .$$

Given $\varepsilon > 0$, choose N such that the second sum is less than ε , and now $\delta > 0$ so small that the first sum is less than ε for $|h| \leq \delta$. Then the right hand side is $\leq 3\varepsilon|h|$ for $|h| \leq \delta$.

Now, let us consider problem (8) in a general Banach space X with a densely defined linear dissipative operator A . Suppose that (8) has a continuously differentiable solution $x(t,x_0)$, for every initial value $x_0 \in D(A)$. Then these solutions are uniquely determined, by Theorem 3.1 with $\omega = 0$. Therefore, we may consider the operators $T(t): D(A) \to D(A)$ defined by $U(t)x_0 = u(t,x_0)$ for $t \geq 0$. We have

(i) $U(0) = I|_{D(A)}$, $U(t+s) = U(t)U(s) = U(s)U(t)$ for $s,t \geq 0$, since the solutions of (8) are uniquely determined.

(ii) $U(t)x$ is continuous in $t \geq 0$, for each $x \in D(A)$.

(iii) $|U(t)x - U(t)y| \leq |x-y|$ for all $x,y \in D(A)$;

compare the proof to Theorem 3.6 . In general, a family of operators $U(t): C \subset X \to C$ with the properties (i) - (iii) on C (instead of D(A)) is called a <u>continuous contraction semigroup on C</u> , also in case the U(t) are nonlinear. To introduce still another useful concept, we notice that our semigroup of solutions is related to A by the equation

$$(9) \qquad \lim_{h \to 0+} \frac{U(h)x - x}{h} = Ax \quad \text{for} \quad x \in D(A) \quad .$$

Given any continuous contraction semigroup $U(t)$ on $C \subset X$, linear or nonlinear, we may associate with this semigroup an operator

$$A_U \colon D(A_U) \subset C \to X$$

by means of (9) , i.e. $D(A_U) = \left\{ x \in C\colon \text{the limit in (9) exists} \right\}$ and $A_U x = \lim$ for $x \in D(A_U)$. This operator is called the infinitesimal generator of $(U(t))_{t \geq 0}$.

Let us still remain in the linear case. Here, one can show that A_U is a densely defined linear operator which is also closed and dissipative. Furthermore

$$(10) \qquad (U(t)x)' = AU(t)x = U(t)Ax \quad \text{for every} \quad x \in D(A) \; .$$

All of these assertions and everything mentioned without proof in the sequel may be found in the books listed at the end of this section. Now the basic existence result is

Theorem 8.4. Let X be a Banach space and $A\colon D(A) \subset X \to X$ a linear operator with $\overline{D(A)} = X$. Then A is the infinitesimal generator of a uniquely determined continuous contraction semigroup of linear operators on $D(A)$ if and only if one of the following equivalent conditions is satisfied.

(i) $(A-\lambda I)^{-1} \in L(X)$ and $|(A-\lambda I)^{-1}| \leq 1/\lambda$ for all $\lambda > 0$.

(ii) $(Ax,x)_- \leq 0$ on $D(A)$ and $A-\lambda I$ is surjective for some $\lambda > 0$.

The condition (i) is due to Hille and Yosida, while (ii) with $\lambda = 1$ has been given in Lumer/Phillips [112] ; it is easy to see that (ii) implies surjectivity of $A-\lambda I$ for every $\lambda > 0$. Clearly, (10) implies that (8) has a unique continuously differentiable solution on $[0,\infty)$ for every $x \in D(A)$ provided A satisfies the conditions of Theorem 8.4 .

The following example is a generalization of Example 8.2 and has been given by Shinderman [161] .

Example 8.3. Let $X = l^p$ with $1 \leq p < \infty$ and consider the countable system

$$x_i' = -a_{ii}x_i - \sum_{j \neq i} a_{ij}x_j \quad \text{for} \quad i \geq 1 \quad ,$$

where $\mathrm{Re}\, a_{ii} \geq \sum_{j \neq i} \max\left\{ |a_{ij}|, |a_{ji}| \right\}$ and $\sum_{j \neq i} |a_{ij}a_{jj}^{-1}| \leq q < 1$ for

$i \geq 1$. In case $p = 1$, let in addition

$$\sum_{i \neq j} |a_{ij} a_{jj}^{-1}| \leq q < 1 \quad \text{for} \quad j \geq 1 .$$

Then the operator A defined by the matrix corresponding to this system on

$$D(A) = \left\{ x \in l^p : \left(\sum_{i \geq 1} |a_{ii} x_i|^p \right)^{1/p} < \infty \right\}$$

satisfies (ii) of Theorem 8.4 as may be seen after some tedious calculations.

Let us also indicate, by means of a simple example, how results like Theorem 8.4 may be applied in the study of timedependent partial differential equations.

Example 8.4. Consider the Cauchy problem for the heat equation

(11) $$u_t = u_{xx} \text{ in } t > 0 , x \in \mathbb{R} ; u(0,x) = \varphi(x) \text{ in } \mathbb{R} .$$

With respect to the variable x we should consider a space of functions $v: \mathbb{R} \to \mathbb{R}$ which are twice differentiable in some sense so that the operator $A = d^2/(dx^2)$ may be applied to v . For example, let us take the Sobolev space (see [197,p.55])

$$W^{2,2}(\mathbb{R}) = \left\{ v: \mathbb{R} \to \mathbb{R}: v, v', v'' \in L^2(\mathbb{R}) \right\} , \quad ' = \frac{d}{dx}$$

with norm $|v| = |v|_2 + |v'|_2 + |v''|_2$, where $|\cdot|_2$ is the usual norm of $L^2(\mathbb{R})$.
Let $X = L^2(\mathbb{R})$. Then $A: D(A) = W^{2,2}(\mathbb{R}) \subset X \to X$ is densely defined. For $\varphi \in D(A)$ we may regard (11) formally as problem (8) with $x_0 = \varphi$. Since $C_0^2(\mathbb{R})$ is dense in $D(A)$ (see [197,p.57]) it is easy to see that A is dissipative, i.e. $(Av,v) \leq 0$ on $D(A)$. Furthermore, $A-\lambda I$ is surjective for $\lambda > 0$. In fact, let $\hat{v}(\xi)$ denote the Fourier transform of $v \in X$ (see [197,p.153]) . Then $Au-\lambda u = f \in X$ implies $-(\xi^2+\lambda)\hat{u} = \hat{f}$, and therefore $u = -F^{-1}((\xi^2+\lambda)^{-1}\hat{f}) \in W^{2,2}(\mathbb{R}) = D(A)$, where F^{-1} denotes the inverse Fourier transform. Hence, Theorem 8.4 applies.
Clearly our very simple example (11) can be solved explicitely, applying the Fourier transform directly to (11) ; this yields the ordinary problem $\hat{u}' = -\xi^2\hat{u}$, $\hat{u}(0) = \hat{\varphi}$ in the Banach space X ; its solution is $\hat{u}(t) = \hat{\varphi}\exp(-\xi^2 t)$, and therefore $u = \gamma * \varphi$, where γ denotes the fundamental solution of $u_t = u_{xx}$.

In many cases we would expect that a solution of a concrete problem like (11) is much more regular than the solution that comes from (8) .

For example we would expect more differentiability in t for (11) since the equation has constant coefficients, more regularity in x if φ is more regular. For specific classes of operators A and initial values x_o such regularity properties may already be obtained for the solutions of (8) , as may be seen in the books listed below under headings like "analytic semigroups" , "applications of fractional powers of A" , etc.

Concerning the nonhomogeneous equation

$$(12) \qquad x' = Ax + b(t) \quad , \quad x(0) = x_o \in D(A)$$

let us assume that A satisfies the hypotheses of Theorem 8.4 . The semigroup U(t) generated by A is then usually written as $U(t) = e^{At}$, in analogy to the bounded case. Now, we would expect that formula (3) of § 1.4 gives the unique solution of (12) , i.e.

$$x(t) = e^{At}x_o + \int_o^t e^{A(t-s)}b(s)ds \quad .$$

It is not hard to see that this is true for instance in case one of the following conditions is satisfied:

"$b(t) \in D(A)$ and the function Ab(t) is continuous" or "b is continuously differentiable"; see e.g. Krein [92,p.135] and notice that we may assume $A^{-1} \in L(X)$ without loss of generality, since the transform $y(t) = x(t)e^{-\lambda t}$ for some $\lambda > 0$ yields the equivalent problem $y' = (A-\lambda I)y + b(t)e^{-\lambda t}$, $y(0) = x_o$, and $(A-\lambda I)^{-1} \in L(X)$ by assumption.

Many results, similar to those for problem (12) , have been established for the timedependent problem (12) , i.e. with A(t) instead of A , in case D(A(t)) does not depend on t . Under reasonable conditions it is possible to construct an evolution operator U(t,s) , identical with $e^{A(t-s)}$ in case $A(t) \equiv A$, such that the solution of (12) may be written as

$$(13) \qquad x(t) = U(t,0)x_o + \int_o^t U(t,s)b(s)ds \quad ;$$

(remember R(t,s) in § 1.4 for bounded A(t) . Let us mention at least a special case of a "classical" result of Kato [84,Theorem 5] which is related to Theorem 8.4 .

Theorem 8.5. Let X be a Banach space and $A(t): D \subset X \to X$, for $t \in J$ with $J = [0,a]$, a family of linear operators such that D is dense in

X and A(t) satisfies (ii) of Theorem 8.4 . Suppose also that $(A(t)-I)(A(0)-I)^{-1}x$ is continuously differentiable in J , for every $x \in X$. Then there exists an evolution operator $U(t,s): D \to D$ for $0 \leq s \leq t \leq a$ such that $|U(t,s)| \leq 1$, $U(t,t) = I$, U is continuous in (t,s) , $U(t,s) = U(t,\tau)U(\tau,s)$ for $s \leq \tau \leq t$, and $U(t,0)x_o$ is the continuously differentiable solution of problem

$$(14) \qquad x' = A(t)x \ , \quad x(0) = x_o$$

for every $x_o \in D$. If $b: J \to D$ is such that $(A(0)-I)b(t)$ is continuous then x(t) given by (13) is the continuously differentiable solution of problem

$$(15) \qquad x' = A(t)x + b(t) \ , \quad x(0) = x_o \ ,$$

for every $x_o \in D$.

Essentially less is known in case the domain of A(t) is variable. Since these remarks have been included for students not familiar with this theory, let us mention some books, and the references given there, where much more details may be found: Chap. 4 of Balakrishnan [6] , Butzer/Behrens [25] , Friedman [66] , Hille/Phillips [77] , Chap. 9 of Kato [82] , Chap. 4 of Krasnoselskii et al. [90] , Krein [92] , Ladas/Lakshmikantham [94] , Chap. 7 of Martin [113] , Chap. 9 and Chap. 14 of Yosida [197] , and the survey article of Nemijtskii/Vainberg/Gusarova [130] .

Let us close this section with some remarks on nonlinear problems. Consider first the semilinear problem

$$(16) \qquad x' = A(t)x + f(t,x) \ , \quad x(0) = x_o \ ,$$

where A(t) is a linear operator such that results like Theorem 8.5 apply to (12) , and assume that f(t,y(t)) , for some class of D-valued functions y , is such that it can be inserted into (13) in place of b(t) . Then we may apply fixed point theorems to the integral equation

$$x(t) = U(t,0)x_o + \int_0^t U(t,s)f(s,x(s))ds$$

to obtain solutions of (16) . In case the linearized problems $x' = A(t)x + f(t,y(t))$, $x(0) = x_o$ are uniquely solvable, we may also use properties of this solution to find a fixed point of the map $y \to$ solution ; cp. the proof of Theorem 2.4 .
Next, let us also mention a basic result for the nonlinear problem (8) with a dissipative operator A , i.e. $(Ax-Ay,x-y)_- \leq 0$ on D(A) . In

Theorem 8.4 for the linear case we had in particular the assumption $R(A-\lambda I) = X \supset D(A)$ for $\lambda > 0$; since A is linear this is equivalent to $R(I-\mu A) = X$ for $\mu > 0$, and it can be shown that the semigroup generated by A is given by

$$(17) \qquad U(t)x = \lim_{n\to\infty} (I - \tfrac{t}{n}A)^{-n}x \quad \text{for} \quad x \in D(A) \text{ and } t \geq 0 \; ;$$

remember the classical formula $e^{at} = \lim_{n\to\infty} (1 - \frac{t}{n}a)^{-n}$ for $a,t \in \mathbb{R}$. Now, Crandall/Liggett [39] have shown, that the limit (17) exists also in case A is nonlinear but dissipative with $D(A) \subset R(I-\mu A)$ in some interval $(0,\mu_0)$. It turns out that U(t) is nonexpansive. Therefore we may extend U(t) to a nonexpansive map from $\overline{D(A)}$ to $\overline{D(A)}$, denoted by U(t) again. With this definition it can be shown that U is a continuous contraction semigroup on $\overline{D(A)}$. For $x \in D(A)$, U(t)x is also locally Lipschitz in $t \geq 0$; hence U(t)x is differentiable a.e. in case X is reflexive. Now suppose that A is also closed. With these conditions on A , Miyadera [123] has shown that $U(t)x \in D(A)$ and

$$(U(t)x)'\big|_{t=t_0} = AU(t_0)x$$

provided U(t)x is differentiable at $t_0 > 0$. Hence, we have

<u>Theorem 8.6</u>. Let X be a reflexive Banach space, $A: D(A) \subset X \to X$ closed, dissipative and such that $R(I-\mu A) \supset D(A)$ in some interval $(0,\mu_0)$. Then problem (8) has the unique solution $U(t)x_0$ given by (17) . Here, a "solution" x is understood to be a locally absolutely continuous function on $[0,\infty)$ with $x(0) = x_0$, $x(t) \in D(A)$ a.e. and $x'(t) = Ax(t)$ a.e. in $[0,\infty)$.

We hope that the reader is now well prepared and interested in learning more about this theory and its applications, e.g. in the following books: Barbu [7] , Brezis [19] , Browder [21] , Cioranescu [33] , Friedman [66] , Martin [113] , Chap. 14 of Yosida [197] , and the references given there. For applications to interesting physical problems we also recommend the notes of Reed [153] and the paper [85] of Kato.

4. Qualitative properties

An extensive study of stability concerning the linear problem

$$(18) \qquad x' = A(t)x + b(t) \quad , \quad x(0) = x_o \quad ,$$

where $A: [0,\infty) \to L(X)$ and $f: [0,\infty) \to X$ are locally Bochner integrable, is represented in Massera/Schäffer [120] ; see also Dalecki/Krein [40]. Some stability results for nonlinear problems may be found e.g. in Ladas/Lakshmikantham [94] , Lakshmikantham [96] . The only references we have found for stability of solutions of general countable systems are those given in § 7 .
The book of Massera and Schäffer contains also some results on Floquet representation and existence of periodic solutions for equation (18) . For existence of periodic solutions of nonlinear problems see for example Browder [22] , Straskraba/Vejvoda [166] , [167] and the references given there. For almost periodic solutions see e.g. Amerio [2] and Taam [170] .

Bibliography

[1] AMBROSETTI, A.: Un teorema di esistenza per le equazioni differenziali negli spazi di Banach. Rend. Sem. Mat. Univ. Padova, 39 , 349-360 (1967)

[2] AMERIO, L. and G. PROUSE: Almost-periodic functions and functional equations. Van Nostrand Reinhold Comp.,NewYork 1971

[3] ARLEY, N.: On the theory of stochastic processes and their applications to the theory of cosmic radiation. Gads Forlag, Copenhagen 1948 (second ed.)

[4] - and V. BORCHSENIUS: On the theory of infinite systems of differential equations and their application to the theory of stochastic processes and the perturbation theory of quantum mechanics. Acta Math. 76, 261-322 (1945)

[5] BAGAUTDINOV, G.N.: On existence of solutions of countable systems of differential equations. Izv. Akad. Nauk Kazah. SSR , ser, fiz. mat. 1966 Nr.3, 10-18 (1966)

[6] BALAKRISHNAN, A.V.: Applied functional analysis. Appl. of Math. Vol. 3 , Springer Verlag 1976

[7] BARBU, V.: Nonlinear semigroups and differential equations in Banach spaces. Noordhoff Int. Publ., Leyden 1976

[8] BELLMAN, R.: The boundedness of solutions of infinite systems of linear differential equations. Duke Math. J. 14 , 695-706 (1947)

[9] - Methods of nonlinear analysis. Vol. II . Press, New York 1973

[10] - and R.M. WILCOX: Truncation and preservation of moment properties for Fokker-Planck moment equations. J. Math. Anal. Appl. 32 , 532-542 (1970)

[11] BHARUCHA-REID, A.T.: Random integral equations. Acad. Press, New York 1972

[12] BINDING, P.: On infinite dimensional equations, J. Diff. Eqs. (to appear)

[13] BISHOP, E. and R.R. PHELPS: The support functionals of a convex set. Proc. Sympos. Pure Math. 7 (convexity) , 27-35 (1963)

[14] BITTNER, L.: Die elementaren Differential- und Integralungleichungen mit einem algemeinen Ungleichungsbegriff. Math. Nachr. 38 , 1-17 (1968)

[15] BONY, J.M.: Principe du maximum, inégalité de Harnack et unicité du probleme de Cauchy pour les opérateurs elliptiques dégenerés. Ann. Inst. Fourier, Grenoble 19 , 277-304 (1969)

[16] BOREL, E.: Ann. de l'Ecole Norm. Sup. (3) , p. 35 ff (1895)

[17] BOURGUIGNON, J.P. and H. BREZIS: Remarks on the Euler equation. J. Funct. Anal. 15 , 341-363 (1974)

[18] BREZIS, H.: On a characterization of flow-invariant sets. Comm. Pure Appl. Math. 23 , 261-263 (1970)

[19] - Operateurs maximaux monotones. Math. Studies Vol. 5 , North-Holland Publ. Comp. 1973

[20] BRILL, H.: A class of semilinear evolution equations in Banach spaces. J. Diff. Eqs. (to appear)

[21] BROWDER, F.E.: Nonlinear operators and nonlinear equations of evolution in Banach spaces. Proc. Sympos. Pure Math. Vol. 18 II, Amer. Math. Soc. 1976

[22] - Periodic solutions of nonlinear equations of evolution in infinite dimensional spaces. Lect. Diff. Eqs. (ed. by K. AZIZ) Vol. 1 , 71-96 . Van Nostrand, New York 1969

[23] - Normal solvability and Fredholm alternative for mappings into infinite dimensional manifolds. J. Funct. Anal. 8 , 250-274 (1971)

[24] - Nonlinear eigenvalue problems and group invariance. p. 1-58 in "Functional analysis and related fields" (F.E. Browder, Ed.), Springer-Verlag 1970

[25] BUTZER, P.L. and H. BEHRENS: Semi-groups of operators and approximation. Springer Verlag 1967

[26] CARLEMAN, T.: Application de la theorie des équations intégrales linéaires aux systèmes d'équations différentielles. Acta Math. 59 , 63-87 (1932)

[27] CELLINA, A.: On the existence of solutions of ordinary differential equations in Banach spaces. Funkc. Ekvac. 14 , 129-136 (1971)

[28] - On the local existence of solutions of ordinary differential equations. Bull. Acad. Polon. Sci., ser. math. astr. et phys. 20 , 293-296 (1972)

[29] - On the nonexistence of solutions of differential equations in nonreflexive spaces. Bull. Amer. Math. Soc. 78 , 1069-1072 (1972)

[30] - and G. PIANIGIANI: On the prolungability of solutions of autonomous differential equations. Boll. Unione Math. Ital. (4) 9 , 824-830 (1974)

[31] CHALON, P. and L. SHAW: An expansion for evaluation of sensitivity with respect to a random parameter. Automatica 5 , 265-273 (1969)

[32] CHOW, S.N. and J.D. SCHUUR: An existence theorem for ordinary differential equations in Banach spaces. Bull. Amer. Math. Soc. 77 , 1018-1020 (1971)

[33] CIORANESCU, I.: Aplicatii de dualitate in analiza functionala neliniara. Edit. Acad. Rep. Soc. Romania, Bucaresti 1974

[34] CODDINGTON, E. and N. LEVINSON: Theory of ordinary differential equations. Mac Graw-Hill, New York 1955

[35] CORDUNEANU, C.: Equazioni differenziali negli spazi di Banach, teoremi di existenza e di prolungabilità. Atti Accad. Naz. Lincei Rend. cl. sci fiz. mat. (8) 23, 226 -230 (1957)

[36] COX, D.R. and H.D. Miller: The theory of stochastic processes. Methuen, London 1965

[37] CRANDALL, M.G.: Differential equations on closed sets. J. Math. Soc. Japan 22 , 443-455 (1970)

[38] - A generalization of Peano's existence theorem and flow invariance. Proc. Amer. Math. Soc. 36 , 151-155 (1972)

[39] - and T.M. LIGETT: Generation of semigroups of nonlinear transformations on general Banach spaces. Amer. J. Math. 93 , 265-298 (1971)

[40] DALECKII, J.L. and M.G. KREIN: Stability of solutions of differential equations in Banach space. Transl. Math. Mono. Vol. 43 , Amer. Math. Soc. 1974

[41] DANES, J.: Fixed point theorems, Nemyckii and Uryson operators and continuity of nonlinear mappings. Comm. Math. Univ. Carol. 11 , 481-500 (1970)

[42] DARBO, G.: Punti uniti in trasformationi a condominio non compatto. Rend. Sem. Mat. Univ. Padova 24 , 84 - 92 (1955)

[43] DAY, M.M.: Normed linear spaces. Springer Verlag 1973 (third ed.)

[44] DE BLASI, F.: The measure of weak non compactness. Report 7 (1974/75) Istit. Mat. "Ulisse Dini" Univ. di Firenze

[45] - and J. MYJAK: Two density properties of ordinary differential equations. Rep. 15 (1975/76) Ist. Mat. Univ. di Firenze

[46] DEIMLING, K.: Nichtlineare Gleichungen und Abbildungsgrade. Springer Verlag 1974

[47] - Fixed points of generalized P-compact operators. Math. Z. 188-196 (1970)

[48] - On approximate solutions of differential equations in Banach spaces. Math. Ann. 212, 79-88 (1974)

[49] - Zeros of accretive operators. Manuscripta Math. 13 , 365-374 (1974)

[50] - On existence and uniqueness for differential equations. Ann. Mat. Pura Appl. (IV), 106 , 1-12 (1975)

[51] - On existence and uniqueness for Cauchy's problem in infinite dimensional Banach spaces. Proc. Colloq. Math. Soc. Janos Bolyai Vol. 15 . Diff. Eqs. 131-142 (1975)

[52] DIAZ, J.B. and J.M. BOWNDS: Euler-Cauchy polygons and the local existence of solutions to abstract ordinary differential equations. Funk. Ekvac. 15 , 193-207 (1972)

[53] DICKEY, R.W.: Infinite systems of nonlinear oscillation equations related to the string. Proc. Amer. Math. Soc. 23 , 459-468 (1969)

[54] - Infinite systems of nonlinear oszillation equations. J. Diff. Eqs. 8 , 16-26 (1970)

[55] DIEUDONNE, J.: Deux examples d'équations différentielles. Acta Sci. Math. (Szeged) 12 B , 38-40 (1950)

[56] DOLPH, C.L. and D.C. LEWIS: On the application of infinite systems of ordinary differential equations to perturbations of plane Poisenille flows. Quart. Appl. Math. 16 , 97-110 (1958)

[57] DUGUNDJI, J.: An extension of Tietze's theorem. Pacific. J. Math. 1 , 353-367 (1951)

[58] DU CHATEAU, P.: The Cauchy-Goursat problem. Memoirs Amer. Math. Soc. Vol. 118 , (1972)

[59] EDWARDS, R.E.: Functional analysis. Holt/Rinehart/Winston , New York 1965

[60] EISENFELD, J. and V. LAKSHMIKANTHAM: On the existence of solutions of differential equations in a Banach space. Techn. Report No 8 , Univ. of Texas at Arlington (1974)

[61] EVANS, J.W. and J.A. FEROE: Successive approximations and the general uniqueness theorem. Amer. J. Math. 96 , 505-510 (1974)

[62] FELLER, W.: An introduction to probability theory and its applications. John Wiley, New York 1968 (third ed.)

[63] - On the integrodifferential equations of purely discontinuous Markoff processes. Trans. Amer. Math. Sic. 48 , 488-515 (1940)

[64] FINE, N.J.: On the Walsh functions. Trans. Amer. Math. Soc. 65 , 372-414 (1949)

[65] FITZGIBBON, W.E.: Weakly continuous nonlinear accretive operators in reflexive Banach spaces. Proc. Amer. Math. Soc. 41 , 229-236 (1973)

[66] FRIEDMAN, A.: Partial differential equations. Holt/Rinehart/ Winston, New York 1969

[67] FUCIK, S. et al.: Spectral analysis of nonlinear operators. Lect. Notes 346 , Springer Verlag 1973

[68] GODUNOV, A.N.: On the theorem of Peano in Banach spaces. Funk. Analiz i Prilog. 9 , 59-60 (1975)

[69] GOEBEL, K. and W. RZYMOWSKI: An existence theorem for the equation x' = f(t,x) in Banach spaces. Bull. Acad. Polon. Sci. 18 , 367-370 (1970)

[70] GOLDSTEIN, J.: Uniqueness for nonlinear Cauchy problems in Banach spaces. Proc. Amer. Math. Soc. 53 , 91-95 (1975)

[71] HART, W.L.: Differential equations and implicite functions in infinitely many variables. Trans. Amer. Math. Soc. 18 , 125-160 (1917)

[72] HARTMAN, P.: On invariant sets and on a theorem of Wazewski. Proc. Amer. Math. Soc. 32 , 511-520 (1972)

[73] HEUSER, H.: Funktional Analysis. Teubner, Stuttgart 1975

[74] HEWITT, E. and K. STROMBERG: Real and abstract analysis. Springer Verlag 1969

[75] HILLE, E.: Remarques sur les systèmes des équations différentielles linéaires à une infinité d'inconnues. J. Math. Pure Appliquées (9) 37 , 375-383 (1958)

[76] - Pathology of infinite systems of linear first order differential equations with constant coefficients. Ann.Mat.Pura Appl. 55 , 133-148 (1961)

([74] and [75] are reprinted in HILLE, E.: "Classical analysis and functional analysis" selected papers, ed, by R.R. Kallman, The MIT-Press, Cambridge 1975)

[77] - and R.S. PHILLIPS: Functional analysis and semigroups. Amer. Math. Soc. Colloq. Publ. Vol. 31 , Providence 1957

[78] HOLMES, R.B.: A course on optimization and best approximation. Lect. Notes 257 , Springer Verlag 1972

[79] HOPF, E.: Über die Anfangswertaufgabe für die hydrodynamischen Grundgleichungen. Math. Nachr. 4 , 213-231 (1951)

[80] JAMES, R.: Reflexivity and the sup of linear functionals. Israel J. Math. 13 , 289-301 (1972)

[81] KARLIN, S.: A first course in stochastic processes. Acad. Press, New York 1965

[82] KATO, T.: Perturbation theory for linear operators. Springer Verlag 1966

[83] - On the semigroups generated by Kolmogoroff's differential equations. J. Math. Soc. Japan 6 , 1-15 (1954)

[84] - Integration of the equation of evolution in a Banach space. J. Math. Soc. Japan 5 , 208-234 (1953)

[85] - Quasi-linear equations of evolution with applications to partial differential equations. pp 25-70 in Lect. Notes 448, Springer-Verlag 1975

[86] KLEE, V.: The support of a convex set in a linear normed space. Duke Math. J. 15 , 767-772 (1948)

[87] KNIGHT, W.J.: Solutions of differential equations in Banach spaces. Duke Math. J. 41 , 437-442 (1974)

[88] KÖTHE, G.: Topological vectorspaces I . Springer Verlag 1969

[89] KOMURA, Y.: Nonlinear semigroups in Hilbert spaces. J. Math. Soc. Japan 19 , 493-507 (1967)

[90] KRASNOSELSKII, M.A. et al.: Integral operators in spaces of summable functions. Noordhoff Int. Publ., Leyden 1976

[91] - and S.G. KREIN: A contribution to the theory of ordinary differential equations in Banach spaces. Trud. Sem. Funkt. Analiz Voronesh Univ. 2 , 3-23 (1956)

[92] KREIN, S.G.: Linear differential equations in Banach spaces. Transl. of Math. Monogr. Vol. 29 , Amer. Math. Soc. , Providence 1971

[93] KURATOWSKI, C.: Sur les espaces complets. Fund. Math. 15 , 301-309 (1930)

[94] LADAS, G.E. and V. LAKSHMIKANTHAM: Differential equations in abstract spaces. Acad. Press, New York 1972

[95] LADDE, G.S. and V. LAKSHMIKANTHAM: On flow-invariant sets. Pacific J. Math. 51 , 215-220 (1974)

[96] LAKSHMIKANTHAM, V.: Stability and asymptotic behaviour of solutions of differential equations in a Banach space. In "Stability problems" Lecture Notes CIME (L. Salvadori, ed.) Ediz. Cremonese, Roma 1974

[97] - and S. LEELA: Differential and integral inequalities, I , II . Acad. Press, New York 1969

[98] - and S. LEELA: On the existence of zeros of Lyapunov-monotone operators. Tech. Rep. No 19 , Univ. of Texas at Arlington (1975)

[99] - and A.R. MITCHELL and R.W. MITCHELL: Maximal and minimal solutions and comparison results for differential equations in abstract cones. Techn. Rep. No 27 , Univ. of Texas at Arlington (1975)

[100] - and A.R. MITCHELL and R.W. MITCHELL: Differential equations on closed sets. Techn. Rep. No 13 , Univ. of Texas at Arlington (1974)

[101] LASOTA, A. and J.A. YORKE: The generic property of existence of solutions of differential equations in Banach spaces. J. Diff. Eqs. 13 , 1-12 (1973)

[102] - and J.A. YORKE: Bounds for periodic solutions of differential equations in Banach spaces. J. Diff. Eqs. 10 , 83-91 (1971)

[103] LEMMERT, R.: Existenz- und Konvergenzsätze für die Prandtlschen Grenzschichtdifferentialgleichungen unter Benutzung der Linienmethode. Thesis, Univ. of Karlsruhe 1974

[104] LEUNG, K.V. and D. MANGERON, M.N. OGUZTÖRELI, R.B. STEIN: On the stability and numerical solutions of two neural models. Utilitas Math. 5 , 167-212 (1974)

[105] LEVINSON, N.: The asymptotic behavior of a system of linear differential equations. Amer. J. Math. 68 , 1-6 (1946)

[106] LEWIS, D.C.: Infinite systems of ordinary differential equations with applications to certain second-order partial differential equations. Trans. Amer. Math. Soc. 34, 792-823 (1933)

[107] LI, T.Y.: Existence of solutions for ordinary differential equations in Banach spaces. J. Diff. Eqs. 18 , 29-40 (1975)

[108] LINDENSTRAUSS, J. and L. TZAFRIRI: Classical Banach spaces. Lect. Notes 338 , Springer Verlag 1973

[109] LOVELADY, D.L. and R.H. MARTIN: A global existence theorem for a nonautonomous differential equation in a Banach space. Proc. Amer. Math. Soc. 35 , 445-449 (1972)

[110] LUDWIG, D.: Stochastic population theories. Lect. Notes Biomath. Vol. 3 Springer Verlag 1974

[111] LUMER, G.: Semi inner product spaces. Trans. Amer. Math. Soc. 100 , 29-43 (1961)

[112] LUMER, G. and R.S. PHILLIPS: Dissipative operators in a Banach space. Pacific J. Math. 11 , 679-698 (1961)

[113] MARTIN, R.H.: Nonlinear operators and differential equations in Banach spaces. John Wiley, New York 1976

[114] - Differential equations on closed subsets of a Banach space. Trans. Amer. Math. Soc. 179 , 399-414 (1973)

[115] - Approximation and existence of solutions to ordinary differential equations in Banach spaces. Funkc. Ekvac. 16 , 195-211 (1973)

[116] - Remarks on ordinary differential equations involving dissipative and compact operators. J. London Math. Soc. (2) 10 , 61-65 (1975)

[117] - A global existence theorem for autonomous differential equations in a Banach space. Proc. Amer. Math. Soc. 26 , 307-314 (1970)

[118] - Lyapunov functions and autonomous differential equations in Banach spaces. Math. Syst. Theory 7 , 66-72 (1973)

[119] - Remarks on differential inequalities in Banach spaces. Proc. Amer. Math. Soc. 53 , 65-71 (1975)

[120] MASSERA, J.L.and J.J. SCHÄFFER: Linear differential equations and function spaces. Acad. Press, New York 1966

[121] MAZUR, S.: Über konvexe Mengen in linearen normierten Räumen. Studia Math. 4 , 70-84 (1933)

[122] McCLURE, J.P. and R. WONG: On infinite systems of linear differential equations. Can. J. Math. 27 , 691-703 (1975)

[123] MIYADERA, I.: Some remarks on semigroups of nonlinear operators. Tohoku Math. J. 23 , 245-258 (1971)

[124] MLAK, W.: Note on maximal solutions of differential equations. Contrib. to differential eqs. Vol. I , 461-465 , Interscience 1963

[125] - and C. OLECH: Integration of infinite systems of differential inequalities. Ann. Polon. Math. 13 , 105-112 (1963)

[126] MOSZYNSKI, K. and A. POKRZYWA: Sur les systèmes infinis d'équations différentielles ordinaires dans certain espaces de Fréchet. Dissertationes Math. 115 , 29 p. , Warszawa 1974

[127] MURAKAMI, H.: On nonlinear ordinary and evolution equations. Funkc. Ekvac. 9 , 151-162 (1966)

[128] NAGUMO, M.: Über die Lage der Integralkurven gewöhnlicher Differentialgleichungen. Proc. Phys.-Math. Soc. Japan 24 , 551-559 (1942)

[129] NAUMANN, J.: Remarks on nonlinear eigenvalue problems. pp 61-84 in "Theory of nonlinear operators" (Proc. Summer School Babylon 1971). Acad. Press, New York 1973

[130] NEMYTSKII, V.V., M.M. VAINBERG and R.S. GUSAROVA: Operational differential equations. in "Progress in Mathematics" Vol. 1 (R.V. GAMKRELIDZE, ed.) , Plenum Press, New York 1968

[131] NIRENBERG, L. and F. TREVES: Local solvability of linear partial differential equations II . Comm. Pure Appl. Math. 23 , 459-510 (1970)

[132] NUSSBAUM, R.: The fixed point index for local condensing maps. Ann. Mat. Pura Appl. 89 , 217-258 (1971)

[133] OGUZTÖRELI, M.N.: On an infinite system of differential equations occurring in the degradations of polymers. Utilitas Math. 1 , 141-155 (1972)

[134] - On the neural equations of Cowan and Stein. Utilitas Math. 2 , 305-317 (1972)

[135] OVCYANNIKOV, L.V.: Singular operators in Banach scales. Dokl. Akad. Nauk SSSR 163 , 819-822 (1965)

[136] PASCALI, D.: Operatori neliniari. Edit. Acad. Rep. Soc. Romania, Bucaresti 1974

[137] PERSIDSKII, K.P.: Countable systems of differential equations and stability of their solutions. Izv. Akad. Nauk Kazach SSR 7 (11) , 52-71 (1959)

[138] - Countable systems of differential equations and stability of their solutions II: The main properties of countable systems of linear differential equations. Izv. Akad. Nauk Kazach SSR 8 (12) , 45-64 (1959)

[139] - Countable systems of differential equations and stability of their solutions III: Fundamental theorems on stability of solutions of countable many differential equations. Izv. Akad. Nauk Kazach SSR 9 (13) , 11-34 (1961)

[140] PHELPS, R.R.: Support cones in Banach spaces and their applications. Advances in Math. 13 , 1-19 (1974)

[141] PHILLIPS, R.S.: Dissipative hyperbolic systems. Trans. Amer. Soc. 86 , 109-173 (1957)

[142] - Dissipative operators and hyperbolic systems of partial differential equations. Trans. Amer. Soc. 90 , 193-254 (1959)

[143] PIANIGIANI, G.: Existence of solutions for ordinary differential equations in Banach spaces. Bull. Acad. Polon. Sci. 23 , 853-857 (1975)

[144] PULVIRENTI, G.: Equazioni differenziali in uno spazio de Banach. Teorema di esistenza e stuttura dell pennello delle soluzioni in ipotesi di Carathéodory. Ann. Mat. Pura Appl. 56 , 281-300 (1961)

[145] RABINOVITZ, P.: Some aspects of nonlinear eigenvalue problems. Rocky Mts. J. Math. 3 , 161-202 (1973)

[146] RAUTMANN, R.: On the convergence of a Galerkin method to solve the initial value problem of a stabilized Navier-Stokes equation. "Numerische Behandlung von Differentialgleichungen". Int. Ser. Num. Math. Vol. 27 , Birkhäuser Verlag, Stuttgart 1975

[147] REDHEFFER, R.: Gewöhnliche Differentialgleichungen mit quasimonotonen Funktionen in normierten linearen Räumen. Arch. Rat. Mech. Anal. 52 , 121-133 (1973)

[148] REDHEFFER, R.: The theorem of Bony and Brezis on flow-invariant sets. Amer. Math. Monthly 79 , 740-747 (1972)

[149] - Matrix differential equations. Bull. Amer. Math. Soc. 81 , 485-488 (1975)

[150] - The sharp maximum principle for nonlinear inequalities. Indiana Univ. Math. J. 21 , 227-248 (1971)

[151] - and W.WALTER: Flow-invariant sets and differential inequalities in normed spaces. Applicable Anal. 5 , 149-161 (1975)

[152] - and W.WALTER: A differential inequality for the distance function in normed linear spaces. Math. Ann. 211 , 299-314 (1974)

[153] REED, M.: Abstract non-linear wave equations. Lect. Notes 507 , Springer Verlag 1976

[154] REID, W.T.: Note on an infinite system of linear differential equations. Ann. Math. 32 , 37-46 (1931)

[155] REUTER, G.E.H.: Denumerable Markov processes and the associated contraction semigroups on l . Acta Math. 97 , 1-46 (1957)

[156] RIESZ, F.: Les systèmes d'équations linéaires à une infinité d'inconnues. Ganthier.Villars, Paris 1913

[157] ROIDER, B.: Walshfunktionen. Jahrbuch Überblicke Mathematik 1976 , BI-Wissenschaftsverlag, Zürich 1976

[158] SADOVSKII, B.N.: Limit-compact and condensing operators. Russ. Math. Surveys 27 , 85-155 (1972)

[159] SHAW, L.: Existence and approximation of solutions to an infinite set of linear timeinvariant differential equations. Siam J. Appl. Math. 22 , 266-279 (1972)

[160] - Solutions for infinite-matrix differential equations. J. Math. Anal. Appl. 41 , 373-383 (1973)

[161] SHINDERMAN, I.D.: Infinite systems of linear differential equations. Diff. Uravnenia 4 , 276-282 (1968)

[162] SHOWALTER, R.E. and T.W.TING: Pseudoparabolic partial differential equations. Siam J. Math. Anal. 1 , 1-26(1970)

[163] SIMHA, R.: J. Appl. Physics 12 , 569 (1941)

[164] STEINBERG, S.: Infinite systems of ordinary differential equations with unbounded coefficients and moment problems. J. Math. Anal. Appl. 41 , 685-694 (1973)

[165] - and F. TREVES: Pseudo Fokker-Planck equations and hyperdifferential operators. J. Diff. Eqs. 8 , 333-366 (1970)

[166] STRASKRABA, I. and O. VEJVODA: Periodic solutions to abstract differential equations. Proc. Equadiff 3 , pp. 199-203 , Brno 1972

[167] - and O. VEJVODA: Periodic solutions to abstract differential equations. Czech. Math. J. 23 , 635-669(1973)

[168] SZUFLA, S.: Measure of noncompactness and ordinary differential equations in Banach spaces. Bull. Acad. Polon. Sci. 19 , 831-835 (1971)

[169] - Structure of the solutions set of ordinary differential equations in Banach spaces. Bull. Acad.

Polon. Sci. 21 , 141-144 (1973)

[170] TAAM. C.T.: Stability, periodicity, and almost periodicity of the solutions of nonlinear differential equations in Banach spaces. J. Math. Mech. 15 , 849-876 (1966)

[171] TREVES, F.: Ovcyannikov theorem and hyperdifferential operators. Notas de Matematica Vol. 46 , IMPA Rio de Janeiro 1968

[172] - Topological vector spaces, distributions and kernels. Acad. Press, New York 1967

[173] TYCHONOV, A.N.: Ein Fixpunktsatz. Math. Ann. 111 , 767-776(1935)

[174] - Über unendliche Systeme von Differentialgleichungen. Mat. Sbornik 41 , 551-555 (1935)

[175] VIDOSSICH, G.: Existence, comparison and asymptotic behavior of solutions of ordinary differential equations in finite and infinite dimensional Banach spaces (to appear)

[176] - Existence, uniqueness and approximation of fixed points as a generic property. Bol. Soc. Brasil. Mat. 5 , 17-29 (1974)

[177] - On the structure of periodic solutions of differential equations. J. Diff. Eqs. 21 , 263-278 (1976)

[178] - How to get zeros of nonlinear operators using the theory of ordinary differential equations. Atas Sem. Anal. Func. Vol. 5 , Sao Paulo, 1973

[179] - On the structure of the set of solutions of nonlinear equations. J. Math. Anal. Appl. 34 , 602-617 (1971)

[180] VOIGT, A.: Line method approximations to the Cauchy problem for nonlinear parabolic differential equations. Numer. Math. 23 , 23-36 (1974)

[181] VOLKMANN, P.: Gewöhnliche Differentialgleichungen mit quasimonoton wachsenden Funktionen in topologischen Vektorräumen. Math. Z. 127 , 157-164 (1972)

[182] - Über die Invarianz konvexer Mengen und Differentialgleichungen in einem normierten Raume. Math. Ann. 203 , 201-210 (1973)

[183] - Über die Invarianzsätze von Bony und Brezis in normierten Räumen. Archiv Math. 26 , 89-93 (1975)

[184] - Über die positive Invarianz einer abgeschlossenen Teilmenge eines Banachschen Raumes bezüglich der Differentialgleichung u' = f(t,u) . J. Reine Angew. Math. 285 , 59-65 (1976)

[185] - New proof of a density theorem for the boundary of a closed set. Proc. Amer.Math.Soc.60, 369-370(1976)

[186] - Über die Existenz von Lösungen der Differentialgleichungen u' = f(u) in einer abgeschlossenen Menge, wenn f eine k-Mengenkontraktion ist. Proc. Conference on ordinary and partial diff. eqs. Dundee 1976, Lecture Notes 564 Springer Verlag 1977

[187] WALTER, W.: Differential and integral inequalities. Springer Verlag 1970

[188] - Ordinary differential inequalities in ordered Banach spaces. J. Diff. Eqs. 9 , 253-261 (1971)

[189] - Gewöhnliche Differential-Ungleichungen im Banachraum. Archiv Math. 20 , 36-47 (1969)

[190] WEINSTEIN, A. and W. STENGER: Methods of intermediate problems for eigenvalues. Acad. Press, New York 1972

[191] WINTNER, A.: Upon a theory of infinite systems of nonlinear implicite and differential equations. Amer. J. Math. 53 , 241-257 (1931)

[192] - Über die Differentialgleichungen der Himmelsmechanik. Math. Ann. 96 , 284 ff (1927)

[193] YORKE, J.A.: Noncontinuable solutions of differential-delay equations. Proc. Amer. Math. Soc. 21 , 648-652 (1969)

[194] - Invariance for ordinary differential equations. Math. System Theory 1 , 353-372 (1967)

[195] - Differential inequalities and non-Lipschitz scalar functions. Math. System Theory 4 , 140-153 (1969)

[196] - A continuous differential equation in Hilbert space without existence. Funkc. Ekvac. 13 , 19-21 (1970)

[197] YOSIDA, K.: Functional analysis. Springer-Verlag 1974 (fourth ed.)

[198] ZAUTYKOV, O.A.: Countable systems of differential equations and their applications. Diff. Uravnenia 1 , 162-170 (1965)

[199] - and K.G. VALEEV: Infinite systems of differential equations. Izdat. "Nauka" Kazach SSR, Alma-Ata 1974

Index